Renewable Energy Policy

Renewable Energy Policy

Paul Komor

Published by the Diebold Institute
for Public Policy Studies
New York

iUniverse, Inc.
New York Lincoln Shanghai

Renewable Energy Policy

iUniverse, Inc.

For information address:
iUniverse, Inc.
2021 Pine Lake Road, Suite 100
Lincoln, NE 68512
www.iuniverse.com

This monograph is part of the Diebold Institute monograph series.

The author is the recipient of a Diebold Institute/Deutsche Bank fellowship to study the use of private capital to create advanced forms of public infrastructure.

ISBN: 0-595-31218-7

Printed in the United States of America

Contents

Acknowledgments

The Diebold Foundation sponsored my year in London to conduct the research for this book, patiently waited for me to complete it, and kindly took responsibility for the publishing. Many thanks to the Foundation and to John Diebold for their generosity and vision.

The Energy Policy and Management Group at Imperial College in London provided all a researcher could ask for—a pleasant place to work, interesting and supportive colleagues, and fine research facilities. I much enjoyed my year there.

Gail Reitenbach and Sonia Smith provided faultless editorial support. Linda Rose provided illustrations. Gunnar Thigpen provided companionship. Patricia Christiansen was a source of inspiration, emotional support, and loving warmth. Peter and Judith Komor provided the tools—intellectual curiosity, dedication to doing good, and a sense of humor. I reserve for myself, however, responsibility for all errors, omissions, and other miscellaneous misdeeds.

Paul Komor
Boulder, Colorado
July 2003

1

Why Renewables? Introduction and Summary

Fossil fuels are the lifeblood of an industrialized society, supplying most of its energy needs. In recent years, the problems with these fossil fuels—including environmental damage, unequal global distribution of fossil fuel resources, price instability, and ultimately supply constraints—have led to a reexamination of their use and a search for alternatives.

Renewable energy—solar, wind, hydropower, and others—is a promising alternative to fossil fuels. Renewable energy is relatively clean, widely available, and the supply is unlimited. Renewable technologies, specifically those for electricity generation, have made impressive technical advances and are now commercially available. Costs have come down considerably and are now close to competitive with fossil fuels. ↑ REALY? LOOK up EXAMPLES.

A closer look at renewables, however, reveals that they too have their limitations. Although costs have come down, renewables are still more expensive in many cases than fossil fuels. Also, not all renewable resources are widely available or evenly distributed; they have their own set of environmental impacts; and some forms of renewables, notably wind, are intermittent and thus a poor match for some electricity systems.

Given these limitations, how do we make the best use of renewables? One approach is through government policy change to promote greater use of renewables. The energy system has long been heavily regulated and publicly controlled. Although this is shifting (see chapter 3), governments still play a major role in energy. Changes in the energy system, therefore, can come about from changes in government policy. But what policies work best? That is the question addressed ✳ in this book.

Specifically, I examine *policies to promote greater use of renewable energy for electricity production*. A number of such policies have been tried, with varying suc-

1

cess. I examine what worked, what didn't, why, and then draw conclusions about what we should do next. This book focuses on renewables for on-grid, large-scale electricity production. *does he look at distributed and small scale electricity generation?*

Problems with Our Current Energy System

The industrialized world's energy system is in many ways an impressive success: It provides dependable light, heat, industrial drivepower, and other basic economic and social needs at a reasonable price. But it has problems as well—problems that are increasingly apparent and that require attention.

Environmental Damage

Fossil fuels provide 80 percent of worldwide energy use (**Table 1-1**).[1] The burning of these fossil fuels contributes to several urgent environmental problems. Climate change (the warming of the earth due to human-induced increases in certain atmospheric gases, notably carbon dioxide [CO_2]) is due largely to fossil fuel burning.[2] Fossil fuel burning also results in emissions of various other pollutants, including sulfur oxides (SO_x), nitrogen oxides (NO_x), and particulates. These pollutants contribute to a variety of public health and environmental concerns. SO_x and NO_x are chief contributors to acid rain as well as local air quality concerns such as smog. Sulfur is found in most coals, although coals vary considerably in their sulfur content. When coal is burned, the sulfur in the coal is emitted into the atmosphere as sulfur dioxide (SO_2) and other sulfate compounds (which as a group are termed SO_x). NO_x is formed when nitrogen in the atmosphere combines with oxygen—a process made possible by the high temperatures of fossil fuel combustion. In the U.S., for example, fossil fuel–fired power plants are responsible for:[3]

- 67 percent of all emissions of SO_2, the leading component of acid rain and fine particulates;

- 25 percent of all emissions of NOx, a key component of smog, acid rain, and fine particulates;

- 34 percent of all emissions of mercury (Hg), a toxic heavy metal that is concentrated through the food chain.

Oil spills, ash waste disposal, mining runoff, and various other environmental problems can also be traced largely to fossil fuel consumption.

Table 1-1: World energy consumption by source, 2000

Source	World energy supply, 2000 (% of total)
Oil	35
Coal	24
Natural gas	21
Biomass/waste[a]	11
Nuclear[b]	7
Hydropower[c]	2
Geothermal, wind, solar	<1
Total	100

Notes:
(a) Includes estimates for "noncommercial" fuels.
(b) Thirty-three percent power plant conversion efficiency assumed.
(c) Converted to energy using direct heat content of electricity (i.e., 3,412 Btu/kilowatt-hour).

Inequitable Distribution of Fossil Fuel Resources

Energy is a universal human need; however, the worldwide distribution of fossil fuels is very uneven. The U.S., for example, is responsible for 26 percent of world annual petroleum consumption but only 12 percent of production.[4] About two-thirds of world crude oil reserves are in the Middle East.[5] Similarly, Asia has 56 percent of the world's population but less than 30 percent of world recoverable coal reserves.[6] This uneven distribution of what has become a necessary resource results in international tensions, trade deficits, and constraints on global development.

Fossil Fuel Resources Are Finite

There is considerable disagreement over just how much oil, natural gas, and coal we have left and when we're going to have to switch to something else. What's *not* in dispute, however, is that these resources are finite. Whether it's in 20 years or 200 years, sooner or later we'll have to find a replacement.

Economic Damage Due to Price Volatility

Oil and natural gas prices are alarmingly volatile. In the U.S., for example, the prices electric utilities paid for natural gas fluctuated between about US$2 to $3 per 1,000 cubic feet ($ft^3$) for most of the late 1980s and 1990s. In 2000, however, gas prices started to climb and reached over $8 per 1,000 ft^3 by December 2000. Prices peaked at $9.47 in January 2001 but, by December 2001, had collapsed down to $3.11.[7] Similarly, the price of crude oil went from $10.60 in January 1999 to $25.60 in January 2000.[8] Such fluctuations cause considerable economic damage and complicate financial planning and forecasting.

wow → to $140 right now

Renewables: Fuel of the Future?

One possible solution to the various problems with our energy system is a switch to renewables. Renewables, broadly speaking, are fuels whose use today does not reduce the supply for tomorrow. Examples are wind energy, solar energy, hydropower, and some forms of biomass. These types of energy are being continually created by the sun, and their future availability is not reduced even if they are diverted for direct human use. Geothermal is often considered renewable as well, although some argue that it is finite and depletable. Less well known forms of renewable energy are tidal power and wave power.

harmful to fish?

Low Environmental Impact

The environmental damage caused by renewables is generally much less than that of fossil fuels. Wind, solar, and hydropower contribute little or no CO_2 to the atmosphere and thus do not directly contribute to climate change.[9] Closed-loop biomass (in which plants are grown specifically for burning) is generally viewed as resulting in no additional carbon emissions, while the carbon impact of open-loop biomass is less clear (it depends on what would have been done with the land or the biomass if it had not been burned). There are no direct air emissions (such as NO_x, CO, SO_x, or particulates) from wind-, solar-, or hydropower-fueled facilities, and thus they do not contribute to local or regional air-quality problems. Wind and solar do not require mining or fuel transport systems, thereby avoiding the associated environmental damage. Although renewables are not entirely innocuous (see discussion of environmental impacts in the next section of this chapter), in general they are much cleaner and less environmentally damaging than fossil fuels.

Nondepletable/Sustainable Resource

Renewable energy is by definition nondepletable, meaning that it cannot be used up, as it is being continually replenished. This is in direct contrast to fossil fuels, whose supply is finite. Similarly, renewable energy is also sustainable, meaning that it could provide for the energy needs of the present without reducing the availability of energy for the future.

Widely Distributed Resource

Renewable resources—wind, sunlight, flowing water, biomass—are much more widely and evenly distributed around the world than are fossil fuels. Photovoltaics, for example, which convert sunlight directly into electricity, will work anywhere the sun shines. This is in sharp contrast to fossil fuels, whose uneven global distribution has contributed to numerous international disputes.

Deep Popular Support

People like renewables. Numerous public polls have shown very strong and consistent public support for renewable energy:

> Trends in public opinion polls have also shown long-standing preferences for renewables and efficiency over other energy sources…the pattern of preferences for using renewables to supply energy has been consistent in the poll data for 20 years. This is one of the strongest patterns identified in all of the [U.S.] national poll data on energy and the environment.[10]

YET IF all purchased, COST↓ = ECON!

Results of public opinion polls have long been criticized for being unrealistic—that is, some have argued that if consumers better understood the costs and other trade-offs involved in energy supply, their support for renewables would fade away. Recent research in the U.S. state of Texas, however, suggests that public support for renewables may be more robust than is commonly believed. Researchers in Texas used Deliberative Polling, a research method used to reduce the influence of poor information on public preferences.[11] A sample of electric utility customers throughout Texas was asked their preferences about a variety of energy topics, including renewables. These customers then underwent in-depth training on energy issues: They were given written material that had been carefully screened for bias; they deliberated the trade-offs in electric generation technologies; and they discussed the issues. The sample was then asked their preferences a second time.

In the first (uninformed) polling, more than 50 percent of interviewees favored renewables as the generation resource that should be pursued first. After the training, which included information on the costs of renewables and other generation options, the preference for renewables dropped moderately to a level roughly tied with energy efficiency—but still ahead of that for fossil fuel plants. In other words, in Texas, consumers still preferred renewables over fossil fuel plants, even when well informed about the costs.[12]

This result is not universal—other areas have seen sharp drop-offs in renewables support once the costs are made clear[13]—but the Texas results do suggest that the public's renewables support is not purely ephemeral or based on ignorance. why texas?

Renewables Have Problems

Although renewables appear to be a promising solution to our energy woes, they too have a number of problems. They are not insurmountable problems, but a realistic view of the potential of renewables requires an honest appraisal of their advantages and their drawbacks.

Renewables Usually Cost More

The costs of renewables have dropped sharply in recent years and, in many cases, will continue to drop. However, at today's costs, they still are far from cheap. **Table 1-2**[14] summarizes current levelized[15] costs and other attributes of renewables. (These costs are discussed in detail in chapter 2.)

Table 1-2: Renewable technology summary

Technology	Typical levelized costs (US cents per kWh)	Advantages	Problems
Wind	4–5	Widespread resource, scalable	Difficult to site, inter-mittent
Photovoltaics	20–40	Ubiquitous resource, silent, long lifetimes	Very expensive, intermittent
Biomass	4–9	Dispatchable, large resource	Has air emissions, expensive
Hydropower	4	Dispatchable, can be inexpensive	Has land, water, and ecological impacts
Geothermal	5–6	Dispatchable, can be inexpensive	Limited resource, depletable

Note: Costs shown are typical for projects built today. Costs will vary widely depending on project specifics.

Whether geothermal and wind, for example, are competitive with fossil fuels is a contentious and unending argument. (Photovoltaics [PVs] clearly cost more.) In terms of first (initial) costs only, the purchase price of a natural gas turbine is on the order of $500 per kilowatt (kW), or roughly half the first costs of a wind turbine. Of course, looking only at first costs is much too simplistic. Pulling operations and maintenance (O&M) and fuel costs into the calculation results in wind-produced electricity being much closer in cost to that of natural gas–fired generation. The specific point at which wind becomes less expensive depends on the prevailing natural gas prices. At natural gas prices of $3 per 1,000 ft³, for example, wind-produced electricity is likely to be close to cost-competitive with electricity from a natural gas turbine. This conclusion will vary depending on assumptions about future natural gas prices, time value of money, and other factors—but in many cases, renewables continue to cost more.

Whether or not renewables are *worth* the extra cost is of course a separate question. As noted earlier, renewables have many advantages over fossil fuels, most notably reduced environmental impact, and energy users may be willing to pay a premium for them.

Renewable Resources Are Not Ubiquitous

Although renewable resources are far more widely and evenly distributed than fossil fuels, they are still uneven. Geothermal resources sufficient for electricity generation, for example, are available only in a few countries. Wind sufficient to drive large wind turbines is more common than many believe but is still geographically limited. Sunlight is omnipresent, but photovoltaics are still impractically expensive. So a switch to renewables would reduce, but not eliminate, the problems associated with access and availability of energy resources.

Renewable Resources Are Intermittent

Intermittency is an issue because of electricity's strange attribute of being very difficult to store. As a result, electricity must be generated on demand. Wind and solar electricity generation is intermittent, meaning that its output is governed not by the system's needs but by the natural fluctuations of the wind and the sun. This reduces the value of the wind and solar output. Geothermal and fossil-fired plants, in contrast, can be operated in response to system needs—at full capacity when electricity needs are highest and then turned down or off when needs are lower.

The penalty, or reduced value, that should be applied to wind and solar due to their intermittency is a contentious issue. At one extreme, some argue that wind and solar should receive no capacity value, that their only value is as a fuel saver. In other words, wind and solar can not be depended on at all, because the wind could stop blowing or the clouds could roll in, and therefore one always will need fossil fuel plants as backup.

At the other end of the argument, some are of the view that a collection of wind turbines located at different locations can be viewed as a reliable power plant, as it is extremely unlikely that the wind will stop blowing everywhere all at once.[16]

The truth, as usual, lies somewhere in between. It's clear, however, that a PV system or a wind turbine is not directly comparable to a natural gas plant with the same rated peak output, and that its output needs to be valued differently.

Renewables Do Have Environmental Impacts

Although renewables are, in general, much less environmentally damaging than fossil fuels, they do have some environmental impacts. Some people see wind turbines, for example, as intrusive and a form of visual pollution. Biomass burning can release significant amounts of particulates and carbon monoxide. Dams built for hydropower inundate large areas of land. All power plants, including renewables, require land for the plant itself as well as transmission lines to transport the electricity. An honest analysis of a renewables project will acknowledge that, although in most cases the environmental impacts are far less than that of a comparable fossil-fired plant, they are not zero.

Where Renewables Are Today

Until the beginning of the twentieth century, renewables in the form of wood were a major source of fuel in the industrialized world. The early twentieth century saw the development of electricity grids in the industrialized world and rapid growth in energy consumption. Most of this new demand was met by fossil fuels: coal, oil, and natural gas. These fuels were preferred over wood—especially for electricity generation—because of their energy density and ease of transport. By the 1930s, fossil fuels were the dominant energy source in the industrialized world, especially for electricity production. As of 2000, renewables, excluding hydropower, provided less than 2 percent of Organisation for Economic Co-operation and Development (OECD) electricity (**Table 1-3**)[17] and less than 3 percent of U.S. electricity (**Table 1-4**).[18]

Table 1-3: Organization for Economic Co-operation and Development (OECD) energy and electricity consumption

Source	OECD energy consumption, 2000 (%)	OECD electricity consumption, 2000 (%)
Oil	41	6
Coal	20	43
Natural gas	22	12
Nuclear	11[a]	32
Biomass/waste	3	1
Hydropower	2[b]	6
Geothermal, wind, solar	<1	<1
Total	100	100

Notes:
(a) Thirty-three percent power plant conversion efficiency assumed.
(b) Converted to energy using direct heat content of electricity
(i.e., 3,412 Btu/kilowatt-hour).

Table 1-4: U.S. energy and electricity consumption

Source	U.S. energy consumption, 2000 (%)	U.S. electricity consumption, 2000 (%)
Oil	39	3
Coal	23	52
Natural gas	25	16
Nuclear	8[a]	20
Biomass/waste	3	2
Hydropower	1[b]	7
Geothermal, wind, solar	<1	<1
Total	100	100

Notes:
(a) Thirty-three percent power plant conversion efficiency assumed.
(b) Converted to energy using direct heat content of electricity
(i.e., 3,412 Btu/kilowatt-hour).

The Case for Policy Intervention

The numerous advantages of renewables make a compelling case for considering options that could increase the use of renewables in electricity generation. Many countries have in fact declared explicitly their desire to see renewables play a greater role. The European Union (EU), for example, has set explicit goals for renewables: 12 percent of energy by 2010.[19] The 2002 World Summit on Sustainable Development concluded with commitments to increase renewables' share of energy consumption.[20] However before jumping into the issue of policy options—that is, how to get from here to there—it's worth laying out the arguments for policy intervention.

A complex web of policies, programs, and subsidies are already in place. One needs a compelling public interest to intervene in a free market. The energy market, however, is about as unfree as a market could be. Electricity production, transmission, and distribution are heavily regulated in most countries (although this is changing; see chapter 3). Energy in all forms is heavily taxed. Most energy technologies, from nuclear to wind, have long received public research and development (R&D) funding. So policies to promote renewables should be seen as policy *change*, rather than new policy *intervention*.

There are numerous environmental externalities associated with fossil fuel exploitation, conversion, and use. These externalities impose costs on others but are not directly reflected in the price of the energy. Although energy taxes may partially address this problem, such taxes may not reflect the societal cost of the externalities. Renewable energy, in contrast, has significantly lower environmental externalities.

Private decisions reflect a short time horizon that ignores long-term problems. It is unrealistic, for example, to expect a private company facing short-term profit pressure to take into account highly uncertain, long-term issues such as climate change in its investment and other decisions. As a result, climate change will generally not impact private sector decisions. As discussed in chapter 3, the private sector's role in energy is growing. Therefore, policy change is needed to ensure that long-term issues such as climate change are reflected in energy technology choices.

It won't happen without policy change. At a more pragmatic level, renewables are unlikely to rise above their paltry current market share without policy change. So if a society wants to see reduced environmental damage, sustainability, and the many other advantages of renewables come to fruition, policies to make it so must be implemented.

Policy vs. markets.
Free economy vs.
gov't regulation.

12   Renewable Energy Policy

Policy and Markets: The Natural Tension

There exists a natural tension between policy and markets—*policy* defined as explicit government actions in pursuit of a specific outcome, and *markets* defined as the functioning of a competitive and lightly regulated or unregulated economic system. The challenge for renewable energy policy is to devise policies that work with the market and intervene as little as possible while still successfully promoting the outcome of increased use of renewables. The ideal choice is not between one or the other, but rather the right spot on the continuum from complete central control to a totally unregulated market.

The focus in this book is on policies that create, support, or otherwise bolster markets for renewable energy. There are three principal reasons for this focus. First, many renewable technologies, notably wind, are beyond the research and development stage. There are diminishing returns to policies that support continued fine-tuning of the technology; what's needed instead is a final nudge to get them in the marketplace. Second, markets create a complex set of forces that help ensure a technology meets the needs of users. This cannot be duplicated by policy. Third, as discussed in chapter 3, there is a global trend of replacing regulation with competition in energy and particularly in electricity supply. If renewables are to succeed, they must succeed in a competitive market.

The Paradox of "Market-Friendly Policies"

The intent of policies discussed here is to help craft a robust, competitive, and thriving market for electricity supply, which includes renewables. This points up a major paradox of renewable energy policy and one that permeates this book. This paradox arises from the inconsistent policy goals of (1) letting markets make technology choice decisions and (2) intervening in the market to ensure that renewables are chosen more often than they would be if the policy didn't exist. That is, "market-friendly policies" have two contradictory goals: letting the market make its own decision, but ensuring that the market decides to use renewables. There is no easy solution to this paradox, but it is worth recognizing, especially when it comes to evaluating renewable policies.

Renewable Technologies

Technologies go through several stages in the long road from conceptual idea to widespread use. The first stage is proof of technology. This stage involves moving from a concept, to more detailed plans, to an actual working model, to a larger-scale test plant, to an operating facility. The next stage is designing to market, which means fine-tuning the engineering and design, tailoring the performance to meet users' needs, and—most important for electricity generation—reducing cost. The third and final stage is market penetration, in which the technology moves from a market-ready idea into widespread use.

Some renewable electricity-generating technologies are still at the proof-of-technology stage: Only a handful of operating wave-power plants exist in the world, for example, and that technology is not yet ready for widespread use. Some technologies, such as hydropower, are widely used and technologically mature. Many, however, are still at the third stage of development, struggling to transition from a proven technology to one in widespread use. These technologies include:

- wind power,
- photovoltaics,
- biomass,
- landfill methane, and
- geothermal.

Selecting the appropriate policy tools to promote a technology requires consideration of its stage of development. Wave power, for example, could benefit from research and development spending, while wind power would see greater benefit from subsidies that reduce its cost to users. Considering policy options to promote renewables, therefore, requires an understanding of just where the various renewable technologies are in terms of cost, technical performance, and other attributes (Table 1-2 above).

Renewable technologies vary widely in costs, resource availability, environmental impacts, and other factors. As a result, it can be misleading to talk about "renewable technologies" without specifying exactly which technology one is talking about. Claims that renewables are prohibitively expensive for many applications or that renewables can act as baseload plants are true only for specific types of renewables (photovoltaics and geothermal, respectively, in this example). Unfortunately, such generic claims are both popular and unhelpful.

It is possible, however, to make some overall conclusions about renewable technologies. The technologies summarized in Table 1-2 are *technically proven*. They are commercially available, reliable, and widely available. These technologies can produce *utility-scale power*. They can be built as multi-megawatt power plants, appropriate for utility generation needs. In addition, for all the technologies except photovoltaics, the *costs are reasonable*. Although renewables typically cost more than fossil-based technologies, the price difference is not huge and can be swamped by renewables' other values, such as reduced environmental impact, wider resource availability, and insulation from fossil-fuel price fluctuations. Therefore, policy discussions about renewables need not dwell on tangential discussions about whether or not the technologies work or if they can produce utility-scale power: They do and they can. Rather, they should address questions of cost, social value, and market efficiency.

Electricity System Restructuring and Renewables

Until the late 1980s, electricity systems in the industrialized world were either government-run or heavily regulated, with new generation investments and retail electricity prices set or overseen by government rather than by market forces. Since then, however, many industrialized countries have restructured ("liberalized") their electricity systems, injecting competition and market forces into the generation and retail parts of the system. The EU has a directive (requirement) that member states open up their electricity systems to competition, and most EU member countries are doing so. Many U.S. states also have introduced some form of competition into their electricity systems. What does this fundamental shift in electricity systems—from regulated monopoly and direct government ownership to competitive, open market—mean for renewables? It creates a plethora of new opportunities, but it closes off a few cherished ones as well.

Overall, the new competitive electricity market has been very good for renewables, creating many new opportunities.

- Competition allows for differentiated retail products such as green power.
- Competition creates incentives for providers to seek competitive advantage, through, for example, branding.
- Along with competition has come new policy tools such as renewable portfolio standards and green certificates.
- A competitive market will have a preference for new generation technologies with shorter construction times.

- Competition suggests the potential for the removal of political borders to electricity, thereby reducing the impacts of unequal renewable resource distribution.

There are, however, a number of ways that restructuring has made life more difficult for renewables:

- Reduced regulatory control makes it harder for governments to use utilities as a tool to promote social goals such as increased use of renewables.

- The new competitive power market has a shorter time horizon and higher discount rate for investments, both of which, all else being equal, reduce the impacts of future fuel costs and thus favor fossil-fueled technologies over renewables.

- Similarly, smaller competitive companies will be more capital-constrained than regulated utilities and thus will prefer technologies with lower capital costs.

- As the wholesale electricity market becomes more liquid and transparent, the intermittent nature of some renewables will become more visible and may reduce the perceived value of these renewables. Similarly, increased use of bilateral contracting may result in reduced prices for intermittent technologies.

Overall, restructuring is moving along briskly in the EU, and although it stumbled mightily in the U.S. due to California's problems of supply shortages and dramatic price increases, it is limping along there as well. Restructuring is neither the death of nor the savior for renewables; it is instead a new environment in which renewables must compete. Renewables are doing well in already-restructured regions, but continuing progress will require jumping on the opportunities restructuring creates, while letting go of the old policy levers as they fade away.

Green Electricity

This book takes a close look at the policy option of "green electricity," programs that allow electricity users to choose between regular "brown" electricity or renewable-based "green" electricity, usually for a 1- to 3-cents-per-kWh premium. We review three countries' experiences with green electricity: the UK, the U.S., and the Netherlands.

Green electricity products in the UK have not been popular. Although green electricity has been available in most of the UK since the advent of retail choice,

BETTER marketing
so more Buy and Prices

as of August 2002, only 60,000 residents had signed up for it—about 0.2 percent of all UK households.[21] This was far behind the sign-up levels in the Netherlands (900,000) and Germany (250,000) at that time. Analysis of the UK green electricity market reveals three principal reasons for its lack of success:

- Restructuring in the UK was presented and defined as a way to reduce costs, thereby making a premium-priced product like green electricity very difficult to sell.

- Green electricity was poorly marketed by the incumbent electricity providers.

- Numerous other policies—notably the climate change levy, the new electricity trading arrangements (NETA), and the renewables obligation—were introduced in the same time period, making energy companies hesitant to invest in renewables due to policy uncertainty.

One analyst noted, "The British green electricity market is in a stalemate of unclear government policies, timid suppliers, and under-enthused consumers."[22] The UK experience, although sobering, makes it clear that a successful green market requires smart marketing from the private sector and policy stability from government.

The U.S. green electricity market has been somewhat more successful, although it too has seen its share of challenges. The first U.S. green power program began in 1993, and since then, more than eighty electricity retailers have entered the U.S. green power business. "It's really pretty basic marketing stuff," reported a utility marketing manager in northern California,[23] whose comment nicely sums up the U.S. experience with green power. Energy retailers that applied essential marketing principles (such as branding and market segmentation) have generally done quite well. A fair number even sold out their green capacity. But others have flopped, either because they were unwilling to invest in the necessary marketing and advertising or because they ran afoul of the shifting regulatory and industry restructuring winds.

The most successful U.S. green power programs have signed up 3 to 6 percent of their residential customers, and some of these have sold out of green energy and so are not accepting new customers. More typical, however, are sign-up levels of about 1 percent. Some green retailers in competitive areas are seeing rapid growth, with the leader in this market, Green Mountain Energy, signing up over half a million customers in seven states. New renewable capacity, mostly wind, is being built to meet this demand. As of February 2003, 980 megawatts (MW) of

how do you sell out?

new renewable capacity had been built expressly to meet the demand for green power in the U.S., with an additional 430 MW planned.[24]

Clearly the green power market is composed of more than an obscure niche of dedicated environmentalists. Success in selling green power to date has been achieved largely through trial and error. Most of the marketing has been done by utilities, which are institutionally unsuited for marketing new products and services. As the understanding of green buyers advances, the marketing will become more sophisticated and successful, and penetration rates will continue to rise. Although the future of the U.S. green power market is very uncertain, one estimate forecasts that an additional 600 to 3,900 MW of new renewable capacity will be built by 2010 to meet green power demand.[25]

The Netherlands' green electricity market is the most successful in the world, with participation rates far above those seen in other countries. This is a remarkable fact given that the Netherlands is a heavily urbanized country with limited renewable energy resources and a ready availability of inexpensive natural gas from its Groningen natural gas fields, which are among the largest natural gas fields in Western Europe.

As of May 2003, green sign-ups in the Netherlands were at 1.8 million—or about 25 percent of the population.[26] To put this in perspective, the more successful U.S. green programs have seen penetration rates of 3 to 6 percent—but this 25 percent Netherlands number is for the *entire country*, not just for a selected program. In fact, some Netherlands green suppliers stopped marketing because they couldn't get enough green electricity to sell.[27] And green electricity sales in the Netherlands are not restricted to the residential sector: In 2001, about one-fourth of Dutch green electricity sales were to nonresidential buyers.[28]

So why is the Netherlands' green market so successful (as measured by market penetration)? There are four principal explanations for its success, listed here in order of importance:

- Due to heavy taxes on fossil-based electricity, green electricity is about the same price as non-green electricity.

- The green market opened first, creating a competitive green market in advance of overall competition.

- Dutch companies have used creative marketing techniques to promote green electricity.

- In addition to fossil fuel taxes, the Dutch government has several other aggressive policies that support renewable electricity.

One obvious reason for the success of the Netherlands' green market is that it's free. If green is about the same price as brown, the argument goes, then of course people will buy it, because they get the green attributes at no cost. A look at other markets, however, shows that this isn't quite accurate. As discussed in Chapter 6, there are several examples of such "free" green electricity products that have not done well. Price is important, certainly, but it's not the entire explanation for the success of the Dutch green market.

The green market in the Netherlands demonstrates that it is possible for green power purchases to achieve a significant market share—the Dutch market is already at 25 percent and is continuing to grow. This was, however, not a "pure" market outcome—the tax exemption and the early opening of the green market to choice had much to do with its success. Key uncertainties right now are the role of imported renewable electricity and the related issue of green certificates. A widely accepted and recognized green certificate trading system in the EU appears to be several years away at best, yet without it there will continue to be controversy over imported renewable electricity.

Overall, a key policy issue is whether the existence of millions of green energy buyers worldwide signifies that renewable energy should be "left to the market"—that is, if consumers want it they will buy it, and it should not be explicitly promoted by policy. This argument is simplistic, however: Consumer decisions to buy green are strongly influenced by price, and the electricity industry and electricity prices are heavily influenced by policy. It's not accurate to characterize consumer decisions to buy green electricity as made in an intervention-free market.

In addition, green buyers are and will likely remain a minority of all electricity buyers. Leaving renewables to green buyers means that renewables will remain a small piece of the electricity pie. If this is a desired outcome, then leaving it to the green market may be appropriate; if not, additional policies need to be considered.

Feed-In Laws

Feed-in laws, which have been used in the U.S. and in Europe, require utilities to buy electricity from renewable generators at high prices, typically at or near the full retail price of electricity. Because these rates are both high and guaranteed by law, investors see little risk in lending for new renewables construction. As a result, investment capital is inexpensive and readily available, and considerable new renewable generation results.

Feed-in laws are in direct opposition to the growing role in the electricity industry of competitive markets and pricing—which are replacing regulation and quotas. The rates paid to renewable generators are seen by many to be unsustainably high, and spreading electricity market liberalization is making it much harder to find a "utility" on which to pin these costs. Yet the past success of feed-in laws means that they have developed a politically powerful constituency that has resisted the phasing out of this heavy-handed policy tool. Additionally, a surprising ruling in 2001 from the European Court of Justice means that reports of the end of feed-in laws, like early reports of Mark Twain's death, were greatly exaggerated.

Feed-in laws are best summarized as "effective but not efficient" (see **Table 1-5**). They do result in considerable new renewable capacity, but at a high price. They are at best a short-term approach, useful at building up an industry from a low level. Care must be taken, however, that they are limited in both scope and duration, for any subsidy creates its own constituency. In Germany's case, this constituency played a key role in ensuring that the subsidy was extended beyond its economic justification.

Table 1-5: Pros and cons of the feed-in law approach

Pros	Cons
Very effective at getting lots of new renewable generation installed	Reduced incentive for cost reduction
Not a direct general-revenue tax	No direct competition between suppliers
Can be very simple	Sets up a dependent and powerful constituency
Costs paid by ratepayers, not general public	Price paid reflects outcome of a political process; does not reflect actual costs
Low revenue uncertainty means low-cost capital	Not a market mechanism—inconsistent with overall European Union direction
Low direct cost to government	Can create stranded costs at a further point in restructuring
Little bureaucratic overhead	Can result in excessive profits for producers

The market response to feed-in laws in both Denmark and Germany shed light on the behavior of the capital markets. Renewable energy is, by its nature, more capital-intensive per kilowatt (kW) of generating capacity than fossil-fired electricity generation. (Renewables have zero fuel costs, of course, so they can still be economically competitive.) This capital intensity often means that the availability of capital limits new renewables construction.

Feed-in laws work largely because the capital markets are happy to lend against the security of a guaranteed-by-law future revenue stream. This is not a realistic feature of a market-based economy. However, it does suggest that innovative ways of reducing the perceived uncertainty of future revenues—for example, by long-term contracts with electricity users or by bounding the price of green certificates—can ensure that renewables appear attractive to the capital markets.

In summary, feed-in laws do yield considerable new renewable capacity, but at high prices. And feed-in laws are a poor fit with the growing role of prices and competitive markets in electricity. The U.S. feed-in law is fading, while Denmark's is in limbo. The combination of entrenched political influence and a favorable court ruling have proven to be more than a match for the restructuring movement in Germany; however, this is likely a short-term result. The overall trend is clear, and feed-in laws do not have a bright future.

The Centralized Bidding System

The UK and Ireland have used a rather complex policy to support renewables: a centralized bidding system combined with a supply-side direct subsidy to generators. The UK's system is called the Non-Fossil Fuel Obligation (NFFO). The NFFO's supply-side subsidy came from an electricity tax, known as the fossil-fuel levy. The portion of the tax revenue from the fossil-fuel levy set aside for renewables was used to cover the difference between the wholesale market price of electricity (which was determined largely by fossil-fired generation) and the cost of renewable electricity. For example, if the prevailing wholesale market price for regular electricity was 3 US cents per kWh, while renewable electricity's wholesale price was 4 US cents per kWh, the regional electricity companies (RECs) paid the renewable generator 3 US cents per kWh, and the electricity tax revenues were used to give the renewable generator the difference of 1 US cent per kWh. The RECs were required to buy the renewable electricity but not to cover its premium costs.

The bidding component of the NFFO came from the process used to determine the price to be paid to the renewable generators. Renewable generators provided a bid—essentially an offer to provide a certain amount of renewable electricity at a certain price—and the government then accepted all bids at or below a cut-off ("strike") price. For each of the five rounds, or bidding cycles, the government set a strike price within a distinct technology "band," or type of renewable generation. For example, for the first-round bidding cycle, the maximum price for wind power was set at 15 US cents per kWh.[29] All offers received by the deadline for wind power at or below that price were accepted, and all that accepted capacity was (in theory) bought by the RECs. To encourage a range of renewable technologies, different types of renewable generation had different maximum prices. For example, for the first round of bidding, the maximum allowable price for landfill gas was set at 9 US cents per kWh[30]—quite a bit lower than wind's 15 US cents per kWh.

The greatest overall success of the NFFO was price reduction. The sharp drops in delivered per-kilowatt-hour costs (for example, wind went from 15 cents per kWh in the first round to 4.3 cents per kWh in the fifth and final round) suggest that the NFFO was successful at getting renewable generation out of the one-off, technology-driven mindset and into the market-based, competitive product world. Although correlation is not necessarily causality—that is, the fact that prices dropped while the NFFO was underway does not necessarily mean that the NFFO *caused* the price drop—other evidence (notably the fact that UK prices for renewables were much lower than that in other countries; see chapter 9) suggests that the NFFO did play a role in driving the price drop.

A subtler but still handy benefit of the NFFO was that its staged implementation—in a series of rounds rather than all at once—allowed for fine-tuning of procedures, prices, allowed technologies, and other details by round.

NFFO had its problems as well. Most of these were not really shortcomings of the NFFO itself, but rather disagreements over what goals the NFFO should have been trying to accomplish. They are generally problems only in the eyes of those whose goals weren't addressed. As the political observation goes, "where you stand depends on where you sit."

Fundamental to any discussion of the NFFO's problems are the inherent contradictions of pursuing a pro-market policy. Almost by definition, a policy is a market intervention intended to accomplish some goal—a goal that presumably would not be met if the policy didn't exist. Many criticisms of the NFFO reveal this tension.

The "picking winners" argument—that the technology bands approach of the NFFO meant that some technologies were favored over others—is a classic argument. It's true, of course, that the NFFO did provide a higher price for wind than for landfill gas, for example. But it's also true that the NFFO "picked" renewables over other electricity generating technologies. NFFO used the technology bands in order to encourage several different renewable technologies, rather than just the least expensive. It's this policy goal, rather than the NFFO itself, that is the point of contention.

Similarly, the NFFO did nothing to further the market readiness of some noncommercial technologies, such as PVs. (Note that this is the flip side of the "picking winners" argument.) Although the NFFO could have been structured with bands for relatively expensive technologies such as PVs, it was not. Here again, the disagreement is over policy goals: Should the NFFO be directed at only the least expensive renewable technology, at a range of close-to-market technologies, or at all renewable technologies? The NFFO took a middle ground and thus was criticized by those at both ends.

Along these same lines, some were disappointed that the NFFO did little to encourage a domestic (UK) renewable technology supply industry. Developers selected the least expensive, most market-ready technology, and in the case of wind, this usually meant a Danish manufacturer. Similarly, some looked to the NFFO to encourage new, small companies to enter the renewables market, but here again NFFO's focus on cost gave an advantage to large, established companies with ready access to lower-cost capital.

NFFO's greatest procedural problem (that is, as distinguished from those related to its choice of goals) was its poor execution rates—that is, its failure to deliver actual, in-the-ground renewable capacity instead of just signed contracts. A ready fix for this—unfortunately at the expense of contract complexity—would be to incorporate a stick of some sort, such as a bond upon contract signing that is returned when actual generation starts.

In the late 1990s, the UK government undertook a series of energy policy reviews. Among the many policy changes that resulted from these reviews was the replacement of the NFFO with a renewable "obligation". It's not accurate, however, to characterize the NFFO as a failure that was eliminated for nonperformance. It certainly had its problems, but it may have been the right tool for its time—a time when many renewable technologies were technically mature but not market-ready. By establishing a guaranteed price for renewable-sourced electricity, it reduced risk to the level where renewable project developers were able to access reasonably priced capital.

Renewable Portfolio Standard (RPS)

A growing number of countries are using a very simple approach to get new renewable capacity built: They are setting a mandatory goal for renewables content and letting the market find the least expensive way to get there. (This book uses the U.S. term *renewable portfolio standard* [RPS] to describe this approach. In the UK it's known as an *obligation,* while other European countries generally use the term *quota*.). Although the names and details vary, the fundamental idea is simple, clear, and—in most cases—quite effective. This approach has wide political support. It gets the nod of approval from both the free market supporters, who like its basis in price, and the renewables advocates who like the certainty of an explicit goal. It has had differing effects on various renewable technologies, however, and this is both its strength and its weakness. In its purest form, it differentiates only on price, which is sometimes not the result that policy makers—and various advocates—want. As with all policies, careful implementation is key.

Most countries have goals for renewables, but the existence of such a goal alone does not make it an RPS. Rather, RPSs are distinguished by:

- Assigning responsibility for meeting the goal to a specific actor—such as electricity users, retailers, or generators.

- Having a substantive penalty for failing to meet the goal.

Most RPSs are based on an explicit annual goal for renewable generation, which can be defined as a percentage of total electricity generation (for example, 10 percent of generation must come from renewable generation by 2010) or as new capacity (for example, 200 MW of new renewables must be added by 2005). These goals usually increase annually. Responsibility for building or buying the required renewable electricity is apportioned in some way—for example, most RPSs require electricity retailers to provide a certain percentage of the electricity they sell as renewable. Most RPSs allow trading, making this policy option closely linked with the idea of green certificates. Some RPS programs allow for "buying out"—essentially setting a cap on costs—while others have a steep penalty for noncompliance.

The RPS is emerging as one of the more popular options for promoting renewable electricity generation. It works better in some situations than others, and its success strongly depends on the many details of its implementation. Like all policy options, it can't be all things to all people—so the more clarity and agreement one can achieve on the policy goals up front, the greater the chances of success.

The RPS's principal advantages include the following:

- *A specified amount of renewable generation is ensured.* A successful RPS will yield the required amount of renewable generation—unlike other policy options, such as financial incentives, whose impacts are often hard to predict.

- *Administrative and bureaucratic costs are low.* Once the goal is set, the only significant government role is monitoring and enforcement. This is typically done through reporting requirements—an annoying but not overwhelming paperwork burden.

- *Price pressure is maintained.* There is no guaranteed price, so (in theory) renewable generators will feel continual market pressure to reduce prices.

- *Risk is reduced, although not eliminated.* An RPS establishes a guaranteed demand for renewable electricity but does not guarantee demand for any specific generator. So overall market risk is reduced but not eliminated—meaning, for example, that reasonably priced capital will be made available.

- *It is simple.* Although there are some interesting subtleties, overall this is one of the simplest and most transparent policy options. This makes it politically attractive.

The RPS approach does have some problems as well:

- *It does not deal well with differing costs across technologies.* A typical RPS has a list of qualifying technologies, but higher-priced ones, such as photovoltaics, will not benefit from the RPS, as they will be unable to compete financially.

- *The renewables goal is set politically and is not price or performance based.* An electricity system uses a mix of generation that is largely determined by price, reliability, and other technical and economic factors. The RPS goal, in contrast, is the outcome of a political process.

- *It is not fundamentally a market mechanism.* The RPS is a regulation and thus imposes some costs.

A number of EU countries and U.S. states already have RPS-like requirements in place. Although many of these have been operating for only a short time, they have already provided lessons about the RPS approach. These lessons include the following:

- *Clarify policy goals.* Is the goal to reduce carbon, to build a domestic renewables industry, to promote fuel diversity? The answer to this question should steer the list of included technologies and other details of the RPS implementation.

- *Set the goal correctly.* The ideal RPS goal will increase the use of renewables to a level higher than what would otherwise occur yet will not unduly increase electricity costs or excessively goad political opponents. Setting the goal needs to be a political process, not an analytical one, and should if possible be the result of consensus.

- *Include technologies carefully.* As discussed in chapter 2, there is not universal agreement on just what qualifies as renewable. Every technology has its advocates, and all will want the RPS to benefit them. Including municipal waste, for example, will engender political support from waste plant developers but will likely result in opposition from environmental groups. The list of included technologies should reflect the policy goal.

- *Expect dissent from higher-priced-technology stakeholders.* Photovoltaics, for example, will not benefit from an RPS unless it contains special provisions to accommodate this more expensive technology.

- *Allow for trading.* This will reduce costs.

Support for the RPS approach is spreading. As of 2003, 15 U.S. states and the UK had RPS-like requirements. Austria, Belgium, Italy, and Sweden had or had considered RPSs. Although the RPS approach is less market-like than, for example, voluntary green markets; its simplicity, transparency, and success at meeting its stated renewable generation goals make it an attractive option.

Green Certificates

Green certificates (also known as tradable renewable certificates [TRCs], green tags, and renewable obligation certificates [ROCs]) are a new and very promising tool for promoting renewable electricity generation. Green certificates are essentially an accounting tool—a way to account for (and monetize) the environmental attributes of renewable-sourced electricity generation. The basic concept behind a green certificate is straightforward. A renewable electricity generator can be thought of as providing two products: the electricity that goes into the grid and the environmental attributes (such as reduced CO_2 relative to fossil-fired generation) associated with the renewable generation. These environmental attributes can be represented by a green certificate. This certificate can then be traded and valued on a secondary market.

Green certificates are a market-based strategy. They do not alter or skew the market, they simply allow it to function more efficiently. As they are not a market intervention or subsidy, they fit neatly into the increasingly pro-market electricity business. They hold the promise of inducing a marked increase in economic efficiency, and the SO_2 trading experience of the U.S. suggests that green certificates could sharply reduce the costs of renewable electricity. When used in conjunction with an RPS or a voluntary green market, they will likely reduce costs and improve market functioning. In addition, green certificates are politically non-controversial, as they require little in the way of direct spending and do not favor one group over another. Finally, their costs are relatively low (relative to direct subsidies or other strategies).

But all is not rosy (see **Table 1-6**). Perhaps most important, the benefits of green certificates are difficult to quantify. Experience to date suggests that they can work well in conjunction with other programs, such as RPSs (as in Texas), but that direct sales of green certificates to end users is problematic due to their complexity and users' lack of familiarity.

Table 1-6: Green certificates' strengths and weaknesses

Strengths	Weaknesses
Widespread political support with little or no direct political opposition	Complicated to understand and implement
	Effects on renewables largely unknown
Generators like them, as they result in a new revenue stream	Unclear relationship with carbon or other pollutant trading
Reasonable administrative costs	International trading raises difficult adminis-
A market mechanism that is economically efficient	trative and policy questions

As of 2003, green certificates were like green energy was in the mid-1990s—a new concept, viewed with a mixture of intrigue and suspicion, that would have to prove itself to an occasionally fickle market. There is currently a flurry of activity at state, national, and international levels to establish working green certificate programs. Although little real-world experience exists with this new and some-what opaque concept, early results suggest that green certificates could play a sig-nificant role in the renewable electricity business.

Close

There is no shortage of interesting and imaginative policy options to promote renewable energy. And there is no shortage of data and opinions—much of it conflicting—about which ones to use when. As this book makes clear, there is no one best option; and each has its strengths and weaknesses.

The search for the optimal mix of policies to promote renewable energy can be simplified considerably by clarifying the goals: Just what are these policies seeking to accomplish? Examples of policy goals might include:

- Reducing environmental harm associated with fossil fuel use,
- Meeting a carbon reduction target at least societal cost.
- Promoting fuel diversity for reasons of national security,
- Minimizing the economic harm of fluctuating fossil fuel prices, and
- Promoting industrial and/or economic development.

Of course, political reality is that all goals have constituencies and all policies are compromises among different, and at times inconsistent, goals. Nevertheless, clarifying goals makes selecting specific policies much easier. For example, adding a "banding" component to an RPS (see chapter 10) increases costs but may also promote industrial development. Prior agreement on goal ranking can help determine if this is a desirable trade-off.

Notes for Chapter 1

1. International Energy Agency, *Key World Energy Statistics* (2002), available at www.iea.org/statist/keyworld2002/keyworld2002.pdf   (downloaded   5 December 2002).

2. In the U.S., for example, "CO_2 from fossil fuel combustion accounted for a nearly constant 79 percent of global warming potential weighted emissions from 1990 to 2000." From U.S. Environmental Protection Agency, "Inventory of U.S. Greenhouse Gas Emissions and Sinks: 1990–2000," EPA-430-R-02-003 (15 April 2002), available at www.epa.gov (downloaded 30 November 2002), p. ES-4.

3. U.S. Environmental Protection Agency, Emissions and Generation Resource Integrated Database, available at www.epa.gov/airmarkets/egrid/factsheet.html (downloaded 30 November 2002).

4. U.S. Department of Energy, Energy Information Administration, "International Energy Annual 2000," DOE/EIA-0219(2000) (May 2002), available at www.eia.doe.gov/iea (downloaded 30 November 2002), pp. 5, 7, 47, 49.

5. U.S. Department of Energy, "International Energy Annual 2000," pp. 112–113.

6. U.S. Department of Energy, "International Energy Annual 2000," pp. 115, 128.

7. U.S. Department of Energy, Energy Information Administration, "Monthly Energy Review," data at www.eia.doe.gov/emeu/mer/prices.html (downloaded 30 November 2002).

8. U.S. Department of Energy, "International Energy Annual 2000," p. 105. Prices here are for Norwegian crude.

9. Some argue that hydropower can contribute to climate change due to methane produced by biological matter in dammed areas.

10. B. Farhar and T. Coburn, "Colorado Homeowner Preferences on Energy and Environmental Policy," NREL/TP-550-25285 (June 1999), available at

www.nrel.gov/docs/fy99osti/25285.pdf (downloaded 30 November 2002), p. 2.

11. For information on this research technique, see www.la.utexas.edu/research/delpol/bluebook/execsum.html.

12. R. Lehr, W. Guild, D. Thomas, and B. Swezey, "Listening to Customers: How Deliberative Polling Helped Build 1,000 MW of New Renewable Energy Projects in Texas," NREL/TP-620-33177 (June 2003).

13. See, for example, "Deliberative Polling," at www.la.utexas.edu/research/delpol/bluebook/delibpoll.html (downloaded 18 June 2003).

14. For sources, see chapter 2.

15. Levelized means including first (capital), operating, maintenance, and fuel costs.

16. C. Archer and M. Jacobson, "Spatial and Temporal Distributions of U.S. Winds and Wind Power at 80 m Derived from Measurements," *Journal of Geophysical Research*, v. 108, no. D9 (2003).

17. International Energy Agency, *Key World Energy Statistics.*

18. U.S. Department of Energy, Energy Information Administration, "Annual Energy Review 2001," DOE/EIA-0384(2001) (November 2002), from www.eia.doe.gov (downloaded 18 June 2003).

19. European Commission, "Energy for the Future: Renewable Sources of Energy," COM(97)599 (26 November 1997), available at http://europa.eu.int/comm/energy/en/com599.htm (downloaded 6 June 2003).

20. See United Nations, "Report of the World Summit on Sustainable Development" (2002), available at www.un.org/jsummit/html/documents/summit_docs.html (downloaded 5 December 2002).

21. Assuming a population of 60 million and an average household size of 2.3.

22. S. Boyle and C. Henderson, "Green Energy: Withering on the Vine?" *Consumer Policy Review* (January/February 2001).

23. P. Komor, "Making Green Electricity Programs Work: The Experts Speak Out," E Source *Green Energy Series* GE-5 (September 2000).

24. L. Bird and B. Swezey, National Renewable Energy Laboratory (NREL), "Estimates of Renewable Energy Developed to Serve Green Power Markets in the United States" (February 2003), from www.eere.energy. gov/greenpower/new_gp_cap.shtmlat (downloaded 22 April 2003).

25. R. Wiser, M. Bolinger, E. Holt, and B. Swezey, "Forecasting the Growth of Green Power Markets in the United States," NREL/TP-620-30101 (October 2001), p. vi.

26. From www.greenprices.com/nl (downloaded 23 April 2003).

27. K. Kwant. and W. Ruijgrok, "Deployment of Renewable Energy in a Liberalized Energy Market," Novem report (undated), from www.novem.org (downloaded 21 June 2002), p. 8.

28. From www.greenprices.com/nl (downloaded 13 April 2001).

29. C. Mitchell, "The England and Wales Non-Fossil Fuel Obligation: History and Lessons," *Annual Review of Energy* (2000), p. 292.

30. C. Mitchell, "England and Wales Non-Fossil Fuel Obligation," p. 292.

2

Renewable Electricity-Generating Technologies: Cost and Performance

Technologies go through several stages in the long road from conceptual idea to widespread use. The first stage is proof of technology. This stage involves moving from a concept, to more detailed plans, to an actual working model, to a larger-scale test plant, to an operating facility. The next stage is designing to market, which means fine-tuning the engineering and design, tailoring the performance to meet users' needs, and—most important for electricity generation—reducing cost. The third and final stage is market penetration, in which the technology moves from a market-ready idea into widespread use.

Some renewable electricity-generating technologies are still at the proof-of-technology stage: Only a handful of operating wave-power plants exist in the world, for example, and that technology is not yet ready for widespread use. Some technologies, such as hydropower, are widely used and technologically mature. Many, however, are still at the third stage of development, struggling to transition from a proven technology to one in widespread use. These technologies include:

- wind power,
- photovoltaics,
- biomass,
- landfill methane, and
- geothermal.

Selecting the appropriate policy tools to promote a technology requires consideration of its stage of development. Wave power, for example, could benefit

31

from research and development spending, while wind power would see greater benefit from subsidies that reduce its cost to users. Considering policy options to promote renewables, therefore, requires an understanding of just where the various renewable technologies are in terms of cost, technical performance, and other attributes.

This chapter assesses a range of renewable technologies, with a focus on real-world cost and performance, current status, and strengths and weaknesses. The focus is on those technologies that have passed the proof-of-technology stage and are market-ready, although perhaps not yet in widespread use. The intent is both to provide a reference for current cost and performance data and to inform later discussions of policy options.

As this chapter shows, renewable technologies vary widely in costs, resource availability, environmental impacts, and other factors (**Table 2-1**). As a result, it can be misleading to talk about "renewable technologies" without specifying exactly which technology is being discussed. For example, one may hear that renewables are prohibitively expensive for many applications. This is true for photovoltaics but not for most other forms of renewables. Similarly, the claim that renewables can act as baseload plants is true for biomass and geothermal but not for solar or wind.

Table 2-1: Renewable technology summary

Technology	Typical levelized costs (US cents per kWh)	Advantages	Problems
Wind	4–5	Widespread resource, scalable	Difficult to site, intermittent
Photovoltaics	20–40	Ubiquitous resource, silent, long lifetimes	Very expensive, intermittent
Biomass	4–9	Dispatchable, large resource	Has air emissions, expensive
Hydropower	4	Dispatchable, can be inexpensive	Has land, water, and ecological impacts
Geothermal	5–6	Dispatchable, can be inexpensive	Limited resource, depletable

Sources: See text.

Wind Power

Wind power is in the midst of a phenomenal boom. In just three years—1999 to 2001—an astonishing 14.2 gigawatts (GW) of wind power was installed world-wide, more than doubling worldwide installed wind capacity.[1] Another 7 GW of wind was added in 2002; by the end of 2002, world wind capacity was at about 31 GW.[2] After many years in which the technical and environmental promise of wind clearly exceeded the commercial reality, wind has turned the corner and is now a commercial, realistic, and even profitable electricity-generating option. Although it is far from trouble-free (it is increasingly hard to site due to land-use conflicts; it is dependent on a fluctuating resource; and its costs are still often higher than those of natural gas–fired generation), it is close to cost-competitive in many areas. And consumers see wind power as very green, making it an espe-cially valuable option for green energy programs.

The Wind Resource

Many people believe that wind turbines require very strong winds. The truth is a moderate but steady wind is much better than a strong gusty one. Wind turbines typically start producing electricity when wind speeds reach 12 to 15 miles per hour (5 to 7 meters per second [m/s]). They reach their rated output at wind speeds of about 30 miles per hour (13 m/s) and typically shut off to avoid damage when wind speeds exceed 50 to 60 miles per hour (22 to 26 m/s). Therefore, any area with sustained wind speeds of greater than 10 to 15 miles per hour may be able to support a wind turbine.

Such sites are surprisingly prevalent. The popular—and incorrect—view is that only a few places have sufficient winds to support turbines. Although it is true that some areas, such as the Great Plains states of the U.S., do have excep-tionally good winds, one analysis found sites with enough wind to justify turbines in forty-five of the fifty U.S. states.[3] A review of European Union (EU) wind resource studies concluded that the EU could get 5 to 10 percent of its electricity from wind, *excluding* any offshore resources.[4] A separate review estimated Europe's "realistic" onshore wind resource at 20 percent of electricity supply.[5]

Offshore wind resources are even greater. There are considerable uncertainties here, but one study found the UK's offshore wind resource to be comparable to the UK's *entire* electricity consumption, even after taking into account shipping, fishing, and other practical constraints. The Scandinavian countries' offshore wind resource is very large and probably exceeds the electricity consumption of those countries as well.[6] A separate study also found Europe's offshore wind resource to exceed its electricity consumption.[7]

There are practical and economic limitations that make much of this resource unrealistic: Specific sites may be too far from transmission lines; they may cause unacceptable visual clutter; power prices may be too low to make such projects financially realistic; and so on. It is, however, important to recognize that the wind resource itself is vast and largely untapped.

Wind Technology

The fundamental idea behind wind power is appealingly simple: The wind turns the blades, the blades turn the generator, and the generator makes electricity. The reality is of course much more complex. Today's wind turbines are technically sophisticated and finely engineered, incorporating the latest advances in materials, microprocessor controls, and computational fluid dynamics (for blade design). The four major components of wind turbines are:

- *The blades.* Today's 2-megawatt (MW) turbines typically have two or three blades, each 130 feet (40 meters) long, made of glass-reinforced plastic or a similar composite material.

- *The nacelle.* The housing on top of the tower contains the generator, gearbox, and related components that convert the rotational energy of the shaft into electricity.

- *The tower.* A typical 2-MW turbine tower is about 250 feet (80 m) high.

- *The balance-of-plant components.* These include roads, foundations, transformers, and various other components.

The wind boom of the late 1990s also saw a number of technical changes. The most dramatic has been the rapid increase in the size (megawatt capacity) of wind turbines. New turbines sold in the mid-1990s typically were rated at about 500 kilowatts (kW), whereas by 2001, half of new turbines sold exceeded 1 MW. Even larger turbines are likely, as prototypes of up to 5 MW are being evaluated.[8]

In the latest generation of turbines, greater attention is being paid to reducing siting problems. The entire assembly is designed to be visually attractive—for example, towers are made of tapered tubular steel, which most people find more visually appealing than the older lattice steel towers. Reduced tip speeds and improved gearboxes all help to keep noise levels at a minimum.

Offshore installations of wind turbines hold great promise as an elegant solution to wind energy's greatest problem: lack of appropriate sites. Although many find wind turbines to be attractive from a distance, few want to look at them every day. Unfortunately, the strongest onshore winds are often found along bluffs near the ocean, at the tops of hills, and along the coast—areas also prized for their physical beauty. The result has been that the more densely settled countries of Europe have found it very difficult to site new wind farms. (This has been less of a problem in the less densely populated western U.S.) Putting the turbines offshore largely solves the siting problem and has the advantage that offshore winds are stronger and more consistent than onshore winds.

Offshore wind turbines do cost more: They require more expensive foundations and underwater transmission cables, and they are harder to access for service and maintenance. However there are savings as well: They don't have the same noise restrictions as onshore turbines, so they can have higher blade tip speeds (which means smaller, less-expensive gearboxes), and they can make use of the better offshore wind resource. It's not yet clear how their overall costs will compare. As of summer 2002, there were firm plans for a 60-MW installation off the coast of Wales and numerous other proposals—including a 100+ MW proposed wind farm in Massachusetts' Nantucket Sound.

Wind Power Costs

How much does wind power cost? This apparently simple question turns out to be surprisingly complex—with much of this complexity due to the fuzziness of the question itself. Sources of uncertainty include:

- *Timing.* Does the question refer to wind turbines now in operation, turbines now being installed, or projected costs at some point in the future?

- *Definition.* Does the question refer to first (initial) costs or costs per unit of electricity (kilowatt-hour) produced? Does the question mean a market price, reflecting any in-place subsidies, or a societal "true" cost?

- *Assumptions.* What is assumed about discount rates, lifetimes, decommissioning costs, and other variables?

- *Wind resource.* What are the wind speeds and durations?

- *Vendors.* Actual costs of the hardware vary by vendor and specific technology used.

Caveats aside, however, a wealth of information exists on wind costs. There are two ways to approach these data. One is to assemble the available information on costs of the various components of wind—turbines, operating costs, and so on—and then combine them to estimate a per-kilowatt (kW) or per-kilowatt-hour (kWh) cost. The other, more direct approach is to simply look at actual wind-power *prices*—that is, to examine contracts for actual wind-generated electricity. Both methods are used here, and the results are very similar: Wind power typically costs 4 to 5 US cents per kWh.

The component approach. Costs for wind include *first* (initial) costs, such as the cost of the wind turbine itself, and *operating* costs, such as maintenance and repair. Each of these cost categories has numerous specific cost items within it. Many of these costs are very site- and project-dependent. For example, if the land is purchased, then the land cost is a first cost, but if the land is leased, then the land cost is an operating cost. Depending on the details of the contract, land lease can be either a fixed annual rent ($/year) or proportional to either capacity (kW) or output (kWh).

For a typical wind project, the *first* costs can be divided into the costs of the turbine itself, which are typically US$650–$700 per kW,[9] and the balance-of-plant costs. These balance-of-plant costs are all the miscellaneous bits and pieces that are needed to complete the project, and include foundation, electrical connection, land-use planning and permitting, land purchase, and other related costs. These are typically up to 40 percent of the turbine costs,[10] or US$270 per kW.[11]

The *operating* costs come in several varieties as well: costs that are proportional to output, such as gearbox maintenance and blade cleaning, and per-year charges, such as insurance and land rent. These costs can all be translated into per-kilowatt-hour costs by making assumptions about annual output. These costs vary by project, of course, but are typically in the range of 0.5 to 0.9 US cents per kWh.[12] Recent estimates are that operating costs for a new turbine installed in 2002 start at about 0.5 US cents per kWh and rise to about 1.0 US cent per kWh after about five years.[13]

Combining these various costs with assumptions about capacity factors, discount rates, and other variables (**Table 2-2**) yields a total generation cost of 4.0 to 4.6 cents per kWh.

Table 2-2: Wind power costs

Type of cost	Cost	Cost in US cents per kWh
Turbine	US$650–$700 per kW	2.5–2.7[a]
Balance of plant	US$270 per kW	1.0[a]
Operating	0.5 to 0.9 US cents per kWh	0.5–0.9
Total		4.0–4.6

Notes: (a) Assuming a capacity factor of 28 percent, a discount rate of 7 percent, a lifetime of twenty years, and no decommissioning costs.
Sources: See text.

The Contract Approach. The alternative view on costs is to simply look at what generators are charging for wind power—that is, actual market prices for wind. These prices reflect technology costs, but also other factors—notably the costs of competing fuels, profits, and various public subsidies.

Although such prices are often treated as confidential for competitive reasons, a few data points are available:

- In the fall of 2001, the state of California signed contracts for 1,800 MW of new wind capacity at an average price of 4.5 US cents per kWh.[14]

- In 1998, the UK contracted for 368 MW of new wind capacity at an average price of 4.2 US cents per kWh.[15]

Wind's Strengths and Weaknesses

Wind's major strength is that its per-kilowatt-hour cost is close to that of new fossil fuel–fired generation. Wind has other attractive features as well:

- It can be sized from about 1 MW to up to hundreds of megawatts, and additional capacity can easily be added in stages.

- It has few adverse local environmental impacts—no emissions, little noise, no waste products—and it is compatible with many land uses, including agriculture and grazing.

- Consumers see wind as very "green" (see chapter 7), and therefore wind-based electricity is a popular green product that can command a healthy premium over regular system power.

It also has a few significant problems, notably:

- The wind resource is limited.
- There is often local opposition to siting of wind farms.
- The electricity production from wind turbines is intermittent.

As discussed above, sites with winds sufficient for electricity generation are more common than many believe, and in some countries, the wind resource exceeds total electricity demand. However, many of these sites are offshore, far from transmission lines, or inaccessible. Even after a site passes these tests, it often runs into siting (planning) conflicts. This problem is especially acute in the more densely settled areas of Western Europe. Many people find the visual impact of wind turbines unacceptable—especially because of the unfortunate overlap of good wind sites and good views. Tops of hills, bluffs along the open ocean, areas unobstructed by topography or large vegetation—these are ideal sites for wind turbines and for sweeping vistas, which don't go together. Local opposition, based on concerns over visual impacts, has squashed many seemingly attractive wind projects. Technical changes have been made in response to siting conflicts, such as slowing tip speeds (which reduce noise) and using solid rather than lattice towers. These have helped, but siting remains a major problem for wind towers. The most promising solution looks to be siting turbines offshore.

The issue of intermittency results from the fact that electricity is very difficult to store. As a result, electricity must be generated on demand. In practice, electric power systems aim to generate just a bit more than is being used, so as not to run short. During peak times, such as summer afternoons when space cooling systems are running, peaking power plants, usually fueled by natural gas, are turned on. Coal-fired and nuclear power plants, in contrast, are baseload plants and operate round-the-clock. Wind turbines, however, generate electricity when the wind blows. Their output cannot be directly controlled but is subject to the whims of the weather. Therefore, they can not operate as either baseload or peaking plants.

There is considerable disagreement over the importance of wind turbines' intermittency. At one extreme, some argue that wind turbines can act only as fuel savers—that, even with wind turbines, enough other power plants need to be operating to meet demand. If the wind does blow, these other plants can be turned down a bit, saving some fuel. Others argue that improved wind forecasting methods make it possible to better predict wind turbine output and that geographically diverse wind turbines considered as one power plant do generate predictable and dependable electricity.[16]

Wind's Future

Wind's combination of relatively low costs (typically 4 to 5 US cents per kWh, as discussed above), environmentally friendly image, and flexibility in sizing means that it has a promising future. Siting will be its single greatest barrier, but if offshore technologies develop as expected, this barrier will be largely removed—adding to wind's bullish future.

Photovoltaics

Photovoltaics (PVs), which convert sunlight directly into electricity, have many attractive features. They are quiet, dependable, have no moving parts, can be installed very quickly, and can be sized to power anything from a single light to an entire community. However, they are quite expensive, with current costs of 20 to 40 US cents per kWh for grid-connected systems (compared to 3 to 5 cents per kWh for coal or natural gas systems). Although costs have come down considerably in recent years and will continue to drop, PVs are currently nowhere near cost-competitive with fossil fuels.

Despite their high costs, PV installations are booming, and worldwide PV production is growing at about 25 percent per year. Although total electricity production from PVs is still quite small relative to coal and other fossil fuels, the use of PVs in various niche markets will mean continued technological and production advances and resulting cost decreases. It will be many years, however, before costs come down to where PVs can compete financially with fossil or wind-based electricity generation.

The Solar Resource

The solar resource is huge and could easily supply the world's electricity needs many times over. For example, PV panels covering 14,000 square miles (36,000 km²)—or less than 15 percent of the land area of Nevada—would provide enough electricity for the entire U.S.[17] (Such a system would be immensely impractical for numerous reasons, notably that it wouldn't generate electricity at night and that it would require massive construction of new transmission lines.) And PVs will work anywhere the sun shines. Although they clearly produce more electricity in sunnier areas, even a cloudy day has enough insolation (sunshine) to produce electricity. Of all the forms of renewable energy, PVs are the least resource-constrained.

PV Technology

PVs convert sunlight directly into electricity with no moving parts, no combustion, no noise, and no waste products.[18] There are two basic types of PV modules: crystalline silicon and thin film. Crystalline silicon modules are used in almost all commercial-scale PV systems—that is, systems producing electricity for resale rather than for direct use. Silicon (which starts out as sand) is mixed with a small amount of a substance with a different number of electrons (such as boron or phosphorus). When light hits the PV material, electrons are dislodged. This movement of electrons creates an electric current. Crystalline silicon modules have reasonably high conversion efficiencies (typically 12 to 14 percent)[19] and are made from readily available materials—usually from waste silicon from chip-manufacturing facilities. Unfortunately, crystalline silicon PV modules are expensive to manufacture.

The newer thin-film PV technology works on the same general principles as crystalline silicon modules but has the advantage of generating electricity from a very thin film. This means the modules could be integrated into building materials (such as roofing tiles). One type of thin film, amorphous silicon, is already commonly used for solar-powered consumer products (such as watches and calculators). In general, thin-film PVs require less material to manufacture than do crystalline silicon PVs and will likely be easier to produce on a large scale. Today's thin-film technologies have lower efficiencies (typically 7 to 10 percent) than do crystalline silicon modules; however, this will likely change in the next five to ten years as thin-film technologies advance.

PV modules (also called panels) typically have a peak power output of 50 to 300 watts. Modules can be assembled into arrays, which can vary from just two modules for a small residential system to thousands of modules for a utility-scale system of 100 kW or more.

The PV modules are the fundamental components of a PV system but certainly not the only ones. Various mounting brackets, supports, and hardware are required to position and hold the modules. An inverter is required to convert the modules' direct current (DC) output to the grid's alternating current (AC) standard. A step-up transformer may be required to increase the voltage to that of the grid. The costs of these nonmodule, or balance-of-system (BOS), components are significant. They typically make up one-third to one-half of total system costs.

How Much Does PV Power Cost?

One sees widely varying costs for PV-sourced electricity. There are several reasons for this variation:

- *How cost is defined.* Calculation of dollars per watt includes only first (initial) costs and does not include operating and maintenance costs. Furthermore, it does not reflect local insolation (sunlight) levels. Calculation of cents per kilowatt-hour, in contrast, does incorporate these factors but also requires assumptions about lifetimes and discount rates.

- *What's included.* A complete utility-scale PV system requires BOS components such as inverters, transformers, and wiring.

- *Whether it's an actual or projected cost.* Costs will continue to decrease with both technical refinements and increased production, and there can be a large difference between what costs might be in a few years and what they really are today.

- *The size and application of the system.* In general, the larger the system, the lower the per-kilowatt and per-kilowatt-hour cost. Utility-scale systems of 100 kW or more are much cheaper, per unit of output, than residential or off-grid systems.

- *Where it's located.* Although PVs will work anywhere, the more sunlight, the lower the per-kilowatt-hour cost.

- *When the system was built.* Recent price drops mean that a system installed today can have much lower costs than one built just a few years ago.

That said, several studies have pulled together and analyzed cost data from actual PV installations:

- An analysis of 220 grid-connected PV systems in the U.S. found average first costs of US$7,400 per kW.[20] Most of these systems were sized for rooftops rather than for utility-scale generation (their average size was 4.9 kW), and most were installed in 1998.

- An analysis of 23 large-scale (70 kW and larger) U.S. PV systems found "costs for large PV systems are dropping and that this trend is likely to continue."[21] Installed PV system costs dropped by 31 percent from an average of US$10.35 per W (US$1,035 per kW) in 1996–1997 to an average of US$8.10 per W (US$8,100 per kW) in 1999–2000. The average cost was US$8,370 per kW. The PV modules themselves accounted for about two-thirds of costs; the remainder was for BOS and installation costs.

- A review of EU cost data estimated PV first costs of US$4,500 to $8,000 per kW in 2000.[22]

Estimating costs per kilowatt-hour brings in additional uncertainties. Operating costs for PV systems are uncertain because there are relatively few in operation. Fuel costs are zero, and scheduled maintenance consists mostly of washing the modules to remove dirt and dust. Technical failures of the modules themselves are very rare. Inverters (which convert the module's DC output to AC) have historically been problematic but are showing improved reliability. One review found maintenance costs for actual grid-tied systems to vary from 0.4 to 9.5 US cents per kWh.[23] A detailed financial analysis of the projected costs of a 2.4-MW PV system estimated operation and maintenance costs at 0.8 US cents per kWh.[24]

So, what do all these costs add up to? Making a number of reasonable but certainly arguable assumptions yields a cost of about 40 US cents per kWh.[25] This is quite expensive—about ten times that of new natural gas turbines, for example. Other published estimates are somewhat lower than 40 cents per kWh, but still much higher than that of most fossil fuel–based technologies. The U.S. Department of Energy reports PV system levelized costs of 20 to 50 US cents per kWh.[26] A separate DOE analysis, which was based on predictions of what it would cost to build large plants rather than on costs from actual installed plants, predicted PV system costs of 18 to 22 US cents per kWh, depending on location.[27]

What Are the Strengths and Weaknesses of PVs?

All electricity sources, renewable and nonrenewable, have strengths and weaknesses. Ideally these factors would be reduced to costs and all combined to yield a final total or "societal" cost. In reality, however, it is often impossible to reduce these factors to financial costs, as their valuation can be subjective and situation dependent. They therefore are discussed here separate from costs.

Strengths. PVs are noiseless and, when placed on rooftops or integrated into building materials, raise few visual hackles. They are therefore amenable to urban siting and less susceptible to siting (planning) disputes than most other electricity sources. They can be sized to fit any application, from a wristwatch to a multi-megawatt utility-scale system.

Although their output will vary depending on the amount of sunlight they receive, they can be installed anywhere the sun shines. Unlike wind turbines, for example, which typically require 12 to 15 miles per hour of wind before they even start turning, PVs will generate some electricity even on a heavily overcast day in winter. The more light, however, the more electricity. So, the question of whether PVs will "work" in a specific geographical location is one of economics and cost-effectiveness, not technical feasibility. For example, Boston, Massachusetts, gets about 40 percent less sunlight than does Albuquerque, New Mexico.[28] A PV system in Boston therefore would have 40 percent fewer kilowatt-hours over which to spread its fixed costs.

Weaknesses. The main problem with PVs is their expense—as discussed above, their cost per unit of electricity output is currently about ten times higher than that of fossil-fired generation. In addition, like some other types of renewable energy, they are intermittent—meaning they operate when the sun shines instead of when they are needed. The degree to which this is a problem depends on the specific application. For example, in an area needing electricity to run space-cooling (air conditioning) systems on sunny days, the coincidence (that is, the match between PV output and the need for electricity) is high, and intermittence is less of a problem. In a region needing electricity at night, however, PVs are much less useful.

PV Applications Today

The PV market is booming, in spite of the high costs. World PV production grew dramatically in the 1990s, from 70 MW per year in 1994 to over 200 MW per year in 1999.[29] The market for this high-priced electricity includes:

- off-grid applications such as road signs, transmission towers, and remote areas;
- consumer electronics such as calculators and watches;
- "green" markets; and
- subsidized and demonstration projects.

This impressive growth rate should, however, be seen in the context of total 1999 PV production of just over 200 MW—which is about the size of just *one* small coal-fired power plant. Put another way, the U.S. total installed PV capacity of 374 MW is just 0.05 percent of the U.S. total electric capacity of 794 GW.[30] Clearly, PVs are not yet a significant source of electricity in the U.S.

PVs' Future

Technical advances in thin-film production, coupled with growing interest in "building-integrated PVs" (BIPV), mean that the PV market will continue to grow rapidly. However, costs will remain uncompetitive with fossil- or wind-based electricity generation for at least five to ten years.

Biomass

Biomass refers to burning biologically produced materials—wood, leaves, various agricultural leftovers, food-processing wastes, and so on—to produce electricity. At present, biomass is the primary energy source for many of the 1.6 billion people in the world who are not served by an electricity grid. In most industrialized countries, in contrast, biomass supplies less than 10 percent of electricity.

The biomass resource far exceeds current use, and numerous technologies are available to convert the varied forms of biomass fuels into electricity. However, biomass' future appears mixed, due to environmental and economic concerns. Biomass burning can result in undesirable air emissions, and the biomass fuel collection and harvesting process has raised questions about its effects on land and soils. Also, in an increasingly market-driven electricity supply business, biomass' relatively high cost per kilowatt-hour has made investors wary.

Biomass: What Is It?

Biomass is the use of biologically produced matter, such as agricultural wastes, trees and plants, and animal wastes, as a fuel. Examples of biomass include:

- wood left over from lumber processing,
- sewage sludge,
- the nonedible portions of crops grown as food,
- dung from farm animals,
- trees and plants collected from land cleared for agriculture, and
- fast-growing trees grown explicitly for use as biomass.

Municipal solid waste is generally not considered biomass, as a significant fraction of it is not biological in origin. Most definitions of biomass do, however, include landfill gas, because it is the decay of biological products that produce the gas. Many of the disagreements over just what should be included as biomass relate to its environmental impacts. For example, the use of post-consumer paper products as a fuel is seen by some as discouraging recycling and thus not environmentally beneficial. As a result, actual definitions of biomass as seen in policy documents and financial support schemes often reflect a political agreement, rather than a technical definition.

The Biomass Resource

The size of the biomass resource reflects how one defines it. The total energy content in plant matter is huge and vastly exceeds human energy needs. Although it would be environmentally damaging and unrealistic to consider most of this as fuel, if one takes a more practical definition of biomass as total crop, wood, and animal wastes, the energy potential is still quite impressive. As shown in **Table 2-3**,[31] existing crop, animal, and wood wastes could supply over one-fifth of Europe's and U.S./Canada's electricity. The potential is even larger if one considers "dedicated" biomass crops (also known as energy crops)—those grown explicitly for use in electricity generation.

Table 2-3: Biomass-waste electricity production potential

Area	Crop wastes (EJ/year)	Animal wastes (EJ/year)	Wood wastes (EJ/year)	Electricity production potential (TWh/yr)[a]	As a percentage of actual 1998 electricity use
U.S. and Canada	5.2	3.1	8.6	940	22
Europe	3.0	4.2	4.4	640	21

Note: EJ is exajoules, or 10^{18} joules.
(a) Assuming 20 percent conversion efficiency. TWh is terawatt-hours.

The potential shown in Table 2-3 is a *technical* potential, meaning that it does not reflect environmental, financial, political, or other concerns which place practical limits on biomass use for electricity generation. For example, although many biomass fuels may be called "wastes," their costs may still be significant due to transportation and processing requirements. Many crop wastes are now used to replenish soils, and their diversion into electricity generation may damage the soil or require expensive (and energy-intensive) chemical fertilizers as replacement. The technical potential for biomass is large, but the practical potential is considerably smaller and continually changing due to changes in technologies, policies, and costs of competing fuels.

Biomass Technologies

It's useful to look at biomass in two separate components: the fuel itself and the technology to convert the fuel to electricity.

The fuel. Most any biological material can be burned for fuel, and in fact much of the world depends on such fuels for cooking and heat. Fuels used for commercial-scale electricity production, however, must be available in large quantities at a reasonable cost. Many of today's biomass fuels are in the form of wood-processing residues, which are the various bits and pieces of wood left over from lumber processing and pulp and paper mills. Second to this is "in-forest residue," which is the tops, limbs, and other pieces left over from forest harvesting. Collecting and transporting these pieces to an electricity generation facility can be expensive. The third major type of biomass fuel now in use is agricultural residues, which includes myriad fuels left over from or associated with agricultural production. Examples include pits, shells, stalks, and prunings. In some cases, electricity production is seen as a solution to a disposal problem: This stuff takes up a lot of room; open burning is highly polluting and usually prohibited; and landfilling is expensive. There are many additional types of biomass fuels, including manure from animals, biogas from anaerobic digestion of biomass, landfill gas (discussed separately below), and sewage sludge.

The electricity conversion technology. Most biomass conversion plants now in operation use direct combustion, meaning that the biomass fuel is burned directly in a boiler, similar to the way coal is typically used. The heat from the burning fuel turns water into steam, and the steam is used to drive a generator. In many applications, the steam is also used for process heat as well—a process called *cogeneration*. For example, many lumber mills in the U.S. use waste wood and scraps as a biomass fuel to produce electricity and steam for use in the mill itself. Extra electricity is sold into the local electricity grid.

The use of conventional steam boilers to convert biomass into electricity has the advantage of using a well-known and familiar technology. Steam boilers have been in use for years and are relatively inexpensive and dependable. One challenge in biomass burning is fuel variability. Biomass fuels, unlike fossil fuels, can vary considerably in moisture content, amount of trace contaminants, energy density (BTUs per pound), and other quantities. Although direct-firing steam systems are relatively tolerant of fuel fluctuations, these fluctuations do make it difficult to control emissions and to operate the plant at maximum efficiency. Steam boilers burning biomass often operate at very low efficiencies, due in part to low fuel quality. One study found biomass-fueled steam systems operating at 14 to 18 percent efficiency—about half that of comparable fossil-fired systems.[32] This low efficiency increases costs and makes biomass less economically competitive.

A number of approaches have been used to improve efficiency and reduce costs. Direct combustion can be improved by improving fuel quality (for example, by drying the fuel and by sorting it to ensure consistency) and by using larger steam turbines with additional cycles to extract more useful heat. An entirely separate approach that does not use direct combustion at all is *gasification*—converting the biomass fuel into a combustible gas, which can then be used to drive a turbine directly. Theoretical efficiencies of such a process appear to be very high, although few such systems are commercially available. A third approach is co-firing in coal-burning power plants, mostly in the 5 to 10 percent range (that is, replacing 5 to 10 percent of the coal with biomass). This would allow the use of biomass in existing coal plants—eliminating the need for a biomass-only turbine. Fuel preparation and handling costs remain a challenge.

What Does Biomass-Sourced Electricity Cost?

Prices of electricity produced from biomass vary quite a bit, due largely to variations in fuel costs. At the low end are electricity production facilities located at industrial sites (such as lumber mills), where the fuel is already there, of known consistency, and essentially free (or at a negative net cost, if burning it avoids disposal costs). At the high end are facilities that must collect or purchase fuel, transport it, and process it before burning. As most biomass fuels are bulky and of relatively low energy density, transport costs quickly become prohibitive outside a radius of 50 to 75 miles (80 to 120 km). This limited fuel supply results in smaller plant capacities. As a result, biomass power plants often produce electricity at a higher cost, as these smaller plants cannot take advantage of the economies of scale that allow 100+ MW plants to produce low-cost electricity.

In general, prices for biomass-sourced electricity are a bit higher than for wind or for most fossil-fired electricity generation—but there is wide variation in prices due to the specific fuels, technologies, and contract details:

- The U.S. Department of Energy estimates that today's biomass plants produce electricity at about 9 US cents per kWh.[33]

- In 2001, the state of California contracted for 3.4 MW of landfill-gas electricity at an average price of 6.1 US cents per kWh, and for 26 MW of biogas from cow dairies at an average price of 7.5 US cents per kWh.

- The UK contracted for a total of 800 MW of electricity from biomass in 1998, at prices that ranged from 3.4 to 3.8 US cents per kWh (**Table 2-4**).[34]

Table 2-4: UK biomass contract prices, 1998

Type of biomass fuel	Capacity (MW)	Average price (US ¢/kWh)
Landfill gas	314	3.8
Municipal/industrial waste	416	3.4
Municipal/industrial waste with CHP[a]	70	3.7

Notes: (a) CHP is combined heat and power, also called cogeneration.

Biomass' Strengths and Weaknesses

As with all forms of electricity generation, biomass has some strengths and weaknesses that may not be fully reflected in price but are nonetheless important.

Strengths. Biomass-based electricity generation has a significant advantage over wind and PV in that it is dispatchable—meaning that it can be turned on and off to meet demand. This makes it a much more valuable resource.

Many of the biomass fuels are bulky, environmentally damaging, and even dangerous. Landfill gas, for example, is extremely flammable and has been the cause of fires and explosions. It is also a potent greenhouse gas, with more than twenty times the global warming potential of carbon dioxide (CO_2).[35] As a result, U.S. regulations *require* landfills to collect and dispose of landfill gas. Sewage sludge dumping in the sea is now prohibited in many areas, and land application is controversial and often illegal. Using these various fuels for electricity generation converts them into (mostly) water and CO_2—vastly reducing the waste-disposal problem.

In some applications, biomass use can actually reduce emissions from fossil-fired plants. Biomass co-firing in coal plants can reduce emissions of nitrogen oxides (NO_x), as a result of the lower nitrogen content of biomass and lower burn temperatures.[36]

Biomass, like all renewable fuels, does have a clear advantage over fossil-fired plants in CO_2 emissions. If the biomass fuel is sustainably harvested—that is, if new vegetation comes up to replace what has been burned—then there are no net CO_2 emissions.

Problems. Biomass burning for electricity generation is not a particularly clean process. Wood burning, for example, can result in significant emissions of particulates. Sulfur (SO_x) emissions from biomass plants are generally lower than that of coal-burning plants[37] but, depending on the fuel, can still be significant. The difficulty with reducing biomass plant emissions is fuel variability—it's very difficult to design and operate a power plant that can efficiently and cleanly burn fuel with varying moisture content, contaminants, and energy density. Some biomass fuels, such as sewage sludge and animal wastes, have various unpleasant and possibly unhealthy contaminants that can be emitted in fuel transport before burning and in the burning itself.

There are subtler environmental issues associated with the diversion of biomass fuels into electricity generation. Crop wastes, for example, are often left in the field and play a crucial role in maintaining soil health. Collecting them for use in electricity generation can damage the soil or require greater use of expensive and energy-intensive chemical fertilizers. Overcollection of in-forest wood can reduce biological diversity and habitat for wildlife.

Biomass' Status Today

For the U.S. and for the EU as a whole, the various forms of biomass supply less than 2 percent of electricity. However, for a few countries, notably Finland and Portugal, biomass is more significant (**Table 2-5**).[38] For many biomass plants, waste disposal and heat production, rather than just electricity generation, are motivations for new construction.

Table 2-5: Electricity generation from biomass, 1998

Country/region	Percentage of electricity generated from...		
	Solid biomass/ animal wastes	Biomass gases and liquids	Municipal waste
EU	1.0	-	-
U.S.	1.0	-	-
Finland	13.3	-	-
The Netherlands	-	-	2.4
Portugal	2.6	-	-
Sweden	1.6	-	-
Switzerland	-	-	1.6

Note: "-" means less than 1 percent.

Biomass' Future

A number of new biomass plants of various types are planned or under construction, but the overall growth appears modest. There is considerable controversy over the extent to which biomass should be included in the various policies and programs now in place and under consideration to promote renewables. For example, only a handful of the 100 or so U.S. voluntary green programs use biomass as a source of green electricity, due largely to concerns that consumers do not view it as sufficiently green. In addition, in contrast to photovoltaics and wind turbines, the costs of biomass-based electricity are not on a steep technology-driven downward curve. Costs are instead dominated by fuel access.

The one area of biomass undergoing relatively rapid growth is biogas (which includes landfill methane, sewage gas, and methane from animal wastes). Many of these projects are driven in part by continuing regulatory and political pressure to reduce odors and flammable gas emissions.

Landfill Methane

Although electricity from landfill gas lacks the visceral appeal of that from solar or wind, it is technically mature and has one distinct environmental advantage: It converts methane (CH_4), a powerful greenhouse gas, into CO_2, which has a much lower per-molecule climate-forcing effect. U.S. regulations implemented in 1996 required larger landfills to install landfill-gas collection systems; this rule gave a considerable boost to this technology. As of 2002, there were over 300 landfill-gas energy systems operating in the U.S. The potential is far from fully tapped: The EPA estimates that there are more than 500 landfills remaining in the U.S. that would be appropriate sites for landfill-methane electricity-generating systems.[39] Such systems can also be found in the UK, the Netherlands, and Germany.

How Does Landfill Methane Work?

Decomposition of organic waste in landfills results in emissions of "landfill gas," which is typically about 50 percent methane. (The rest is mostly CO_2, plus traces of volatile organic compounds [VOCs] and other gases.) Uncontrolled emissions of landfill gas cause several problems:

- *Explosions*. Higher concentrations are explosive, and a number of explosions and fires have been attributed to landfill gas.

- *Foul odors.* It can result in odor complaints from neighbors (the bad smell is mostly due to trace amounts of hydrogen sulfide [H_2S]).

- *Air pollution.* It contributes to local air-pollution problems, notably ozone formation.

- *Global warming.* Methane is a potent greenhouse gas, with twenty-one times the global warming potential, per ton, of CO_2.

Burning landfill gas in a turbine or an engine converts the bulk of the landfill gas into CO_2 and water, thereby solving most of these problems. A landfill-gas electricity system has three components: gas collection, gas treatment, and energy recovery.

Gas collection. Gas is collected via perforated pipes that are sunk into the landfill to depths of 25 to 75 feet (8 to 22 m). A pump applies a vacuum to the pipes, pulling the gas from the landfill into and up the pipes. Condensate is trapped and either sent back into the landfill or pumped to a sewage treatment facility.

Gas treatment. Depending on the final use of the gas, various types of treatment are needed. For electricity generation in an engine, remaining condensate and various impurities must be extracted or filtered out. The gas usually is compressed as well.

Energy recovery. Once the gas is collected, cleaned, and compressed, there are three principal ways it can be used:

- electricity generation via engines, turbines, or other technologies;

- direct use in a boiler to produce hot water or steam or for industrial processes; or

- injection into a natural gas pipeline (this requires CO2 removal and high compression).

The first option, electricity generation, is the most common—about 70 percent of U.S. landfill-gas energy projects generate electricity. Most do so using internal combustion engines, as they are relatively low cost, offer reasonable operating efficiencies, and are an established and proven technology. Disadvantages of using engines include NO_x emissions and susceptibility to corrosion due to impurities in the gas. Engine sizes typically range from 250 kW to 1 MW, and multiple engines can be used for larger projects.

Larger systems can use other generating technologies. Gas (combustion) turbines are feasible for projects over 3 to 4 MW. Systems larger than about 8 MW can use steam turbines. Depending on local demands for steam or hot water, cogeneration may be viable as well.

The Landfill Methane Resource

The two primary rules to determine if a landfill might be a good candidate for a methane-to-electricity system relate to size and age. In general, a landfill must contain at least 1 million tons of waste to be considered a good candidate and must be either operating or recently closed (methane production tails off rapidly after a few years without new waste). After that, questions of climate and waste type come into play: Areas with at least moderate rainfall (more than 25 inches per year) and landfills with organic waste (rather than construction debris, hazardous wastes, or industrial wastes) are better candidates. If a site survives this level of screening, questions of access to transmission lines, local demands for steam or hot water (for cogeneration systems), and expected prices for the electricity produced must be considered. A related issue in considering landfill sites in the U.S. is that federal law requires that landfills with a design capacity over 2.75 million tons install a landfill-gas recovery system.

How Much Does Landfill Methane Cost?

Typical costs for landfill-methane projects are in the range of 6 to 9 US cents per kWh.[40] As with other renewable generating technologies, costs are highly variable depending on the specifics of the project. Key variables include:

- gas productivity (that is, the volume and heat content of gas produced per ton of waste in the landfill),

- whether or not there is a nearby market for steam or hot water (making cogeneration a possibility),

- the availability of municipal financing (which is less expensive than private financing), and

- the total project size (in general, the larger the project, the lower the cost per kilowatt-hour).

Hydropower

Hydropower currently supplies more electricity than most other forms of renewable energy in many countries. Sweden, Switzerland, and Norway, for example, all get the bulk of their power from hydro. Hydropower supplies about 10 percent of all kilowatt-hours in the U.S.—far more than other renewable sources. And electricity from existing hydropower plants is often less expensive than that from either fossil-fired or renewable sources. However, the potential for new electricity generation from hydropower is quite limited, because of environmental effects and resource limitations.

Hydropower is currently a very controversial form of electricity due to concerns over its environmental effects; an increasing number of renewable policies are explicitly limiting its role in electricity systems. For example, California's recently passed renewable portfolio standard excludes new hydropower, and most green pricing programs do not include large hydropower.

How Does Hydropower Work?

It couldn't be simpler. Water—by virtue of its flow rate, elevation ("head"), and pressure—has potential energy. A simple turbine converts this potential energy into electricity. There is no combustion, no heat, no waste products, and almost no noise. The water is not consumed or polluted. A typical hydropower system consists of:

- a dam,
- a water reservoir behind (upstream of) the dam,
- the plumbing ("penstock") that carries the water from the reservoir to the turbine, and
- the turbine and generator.

How Much Does Hydropower Cost?

Initial (capital) costs for recently constructed hydropower plants in the U.S. ranged from US$1,900 to $2,600 per kW of installed capacity.[41] Operation and maintenance are typically about 0.8 US cents per kWh.[42] Assuming a 50-year life, a 45 percent capacity factor, and a 5 percent cost of capital, a total levelized cost would be 3.9 US cents per kWh—lower than most fossil fuel–based and renewable technologies. Note that these costs are for larger systems (with an average size of 31 MW). All else being equal, smaller systems generally have higher costs.

Hydropower's Environmental Effects

Construction and operation of a hydropower plant causes numerous environmental impacts. Careful design and operation can minimize (but not eliminate) these impacts. Furthermore, for some of these impacts, viewing them as "good" or "bad" reflects values, not science. For example, sport anglers may value the increased smallmouth bass population that results from damming a river, while fly fishermen may lament the loss of trout habitat resulting from the same dam. Here we summarize just a few of these impacts.

Ecosystem changes. Building reservoirs causes dramatic changes to the ecosystem both in the reservoir and downstream. The water becomes warmer at the surface and more stratified, and it often has a lower oxygen content. These changes are good for some species and bad for others (for example, smallmouth bass thrive and salmon wane). Large fluctuations in water levels (due, for example, to peak electric demands) make it difficult for the shoreline area to support life. Sediments can build up in the reservoir, further altering water chemistry both in the reservoir and downstream.

Fish passage. Dams make it difficult or impossible for fish to swim the river. Although considerable efforts have been made to devise ways for fish to be able to get through dams, such as through fish ladders, they have not been overwhelmingly successful. This problem is especially acute for salmon in the Pacific Northwest.

Wildlife. The shoreline of a free-flowing river is a biologically active area, supporting numerous plants and animals. Damage to this area can have repercussions for wildlife that depend on the shoreline for food, cover, and other needs.

Hydropower's Future

The technical and resource potential for new hydropower is significant (as it is for most types of renewables). For example, the technical potential for new hydropower in the U.S. is estimated at more than 30 GW.[43] However, most of this 30 GW will never be put to use, because doing so would be seen as environmentally damaging and unacceptable. In fact, the industrialized world will see few if any new large-scale hydropower facilities, due largely to environmental concerns. Iceland's plans for a large new hydropower facility to power aluminum smelters, for example, has proven to be extremely controversial and, as of 2002, was looking uncertain. (Such projects, however, are continuing in the developing world.) That's not to say that there will be *no* new hydropower plants—there are countless small- and medium-size new projects under construction and planned in the industrialized countries. One database identified almost 600 MW of new hydro-power facilities planned in the U.S.[44] And there are ambitious plans in Norway and elsewhere to upgrade and modernize existing hydropower facilities.

Geothermal Electricity

Geothermal electricity has one compelling advantage over wind and solar: It's not limited by the whims of the wind or the sun and can therefore provide baseload electricity. Typical geothermal plants operate at capacity factors of 90 percent, compared to wind's 25 to 30 percent. (*Capacity factor* is defined as actual plant output, divided by theoretical maximum output if the plant operated at full capacity.) This means that a geothermal power plant is delivering close to its maximum output most of the time. This makes it a much more valuable and dependable source of electricity. Geothermal is also technically reliable, environ-mentally clean (although somewhat less so than wind or solar, due to water treat-ment and disposal issues), and usually easier to site than wind.

Geothermal's principal disadvantage is that it is a geographically limited resource. In locations where the natural heat of the earth is found near the earth's surface, it works quite well. Unfortunately, such locations are quite limited. Sites with enough hot water and/or steam to support reasonably priced electricity gen-eration are rare and hard to find. Other constraints include relatively high costs (a bit higher than wind, on average) and limited room for cost reductions. Unlike wind and solar, geothermal is no longer seeing steep cost reductions from tech-nology improvements.

A definitional note: Much of what one reads and hears about geothermal energy is about "direct-use" geothermal—using geothermal hot water and/or steam directly for space heating, water heating, and industrial processes. And ground-source heat pumps are increasingly called geothermal heat pumps, as they also use the heat of the earth to provide useful energy. This discussion, however, focuses on geothermal *electricity*—using geothermal hot water and/or steam to produce electricity.

Geothermal Resources: Where and How Much?

The geothermal resource challenge is finding water or steam that is both close to the surface and hot enough to make it worth exploiting. There are a number of locations in the world with water that is accessible and hot enough to support direct use. Geothermal resources that can support *electricity* generation, however, are much more rare. Only a handful of locations in the world currently are using geothermal resources for electricity generation (**Table 2-6**).[45]

Table 2-6: Installed geothermal electricity-generating capacity

Country	MW installed, 2000
U.S.	2,228
Philippines	1,909
Italy	785
Mexico	755
Indonesia	590
Japan	547
New Zealand	437
Iceland	170
El Salvador	161
Costa Rica	143

Note: Only countries with 100+ MW of geothermal capacity are shown.

Geothermal electricity generation requires large quantities of steam or water that is at least 100 degrees Celsius (and preferably much hotter), close to the surface, and not easily tapped out. Locating these resources is similar to oil and gas exploration: It's possible to locate areas that are promising by examining the local geological conditions, but ultimately one has to dig an exploratory well to see what's really down there. Even then, it's not clear how long that resource will last. (Output at the world's largest geothermal complex, the Geysers in northern California, has dropped sharply due to geothermal resource depletion.) Ultimately, the decision of whether a specific site can support electricity generation is one of economics, educated guesswork, and risk preference. There is a bit of a chicken-and-egg problem here: There has been relatively little geothermal resource exploration in recent years as new geothermal electric capacity has not been cost-effective; but, of course, without such exploration the cost-effective resources won't be found.

There is little doubt that a *technical* potential for increased geothermal electricity in some areas remains. One estimate puts the known, accessible potential in the western U.S. at over 20 GW.[46] Whether or not any of this potential makes *financial* sense is not clear and depends in large part on prevailing power prices.

Geothermal Technology

Most currently operating geothermal electricity plants are either steam or binary cycle. If the geothermal temperatures are sufficiently hot, dry steam from the earth can be used directly to drive a steam turbine—such as is done at the Geysers geothermal complex in northern California. More typically, hot water is flashed into steam by reducing the ambient pressure, and the resulting steam is used to drive a turbine. This is a relatively simple process (at least relative to a binary cycle process), but requires high-temperature water (above 150 degrees C). Solids and other contaminants in the steam can also cause environmental problems.

Lower-temperature resources (100 to 150 degrees C) require a binary cycle system, in which hot brine (a mix of water and various minerals) is pumped from geothermal wells. The heat of the brine is extracted with a heat exchanger, and the cooled brine is then reinjected into the earth. This is a more expensive and complicated process.

A new approach to geothermal—hot dry rock—is intriguing but still far from commercially viable. The idea is to tap the vast amount of heat contained in deep rock formations that do not contain water or steam. The technical challenges are many, including drilling down to these hot dry rocks, injecting and then recovering some sort of heat transfer fluid (probably water), and fracturing the rocks to allow the fluid to penetrate. Although the technical potential is large, the high costs of deep drilling and the technical challenges will keep this technology uneconomic for some time.

Most existing geothermal power plants are over 5 MW. However, small (usually defined as less than 5 MW) geothermal power plants are actively being researched for both on-grid and off-grid applications. Such plants could be located at sites that could also make use of direct heat, increasing the overall cost-effectiveness. In general, however, electricity costs for small geothermal plants will be quite high, and therefore such plants will make economic sense only in applications where the direct heat is of great value.

Geothermal Costs

There are two ways to look at the costs of geothermal power plants: by combining the various component technology costs to come up with an overall *cost* per kilowatt or per kilowatt-hour, or by looking at the actual market *price* for geothermal electricity. The first method has the advantages of revealing the individual component costs and is not swayed by short-term market fluctuations. It is, however, somewhat naive, in that real-world pricing generally reflects market conditions rather than just the cost of the underlying technology. The second approach is a better reflection of reality but is more volatile and strongly influenced by factors not directly related to geothermal—notably fossil fuel prices and investor risk preferences. Both types of data are presented here.

The first (initial) costs of geothermal power plants vary considerably, due mostly to the geothermal resource. The depth of the well, the length of the piping between the well and the turbine, the level of contaminants in the water, the water temperature, access to transmission lines—all these vary and strongly influence the first costs. A useful *rough* estimate for a flashed-steam geothermal power-plant system is about US$1,500 to $2,000 per kW for a larger (5+ MW) system, with these costs roughly split between the power plant itself and the resource infrastructure (well, piping, water treatment, and so on). Binary plants are somewhat more expensive, with system first costs of US$2,000 to $2,500 per kW.[47]

Although there are no fuel costs per se for geothermal plants, there are a number of operation and maintenance (O&M) costs. Plant operators may need to pay some kind of royalty or lease to the landowner, which can be an annual or per-kilowatt-hour cost. The geothermal water or steam may need to be treated to remove minerals and other contaminants. Pumps may be needed to reinject water into the ground. These costs vary as well, but a reasonable assumption for O&M costs is 2 to 3 US cents per kWh.[48]

Combining these first and O&M costs with reasonable assumptions (15-year lifetime, 7.5 percent discount rate, and 89 percent plant capacity factor) yields an amortized energy cost (also called levelized energy cost) of 5.0 US cents per kWh for a flashed-steam plant and 5.8 US cents per kWh for a binary plant. With an ideal geothermal resource (very hot water or steam close to the surface, few contaminants, power plant located very close to the well, good access to transmission lines, and so on), one could do better than this: The Geysers plant in northern California, one of the largest geothermal facilities in the world, sells its power at 3 to 3.5 US cents per kWh.[49]

One can also look at geothermal costs from the perspective of actual signed contracts. As noted earlier, this is a price rather than an engineering- or technology-based cost estimate. It reflects market conditions (notably the prices of competing fuels) at the time the contract was signed and thus may not be representative of prices under different market conditions. That said, it's useful to note that in 2001 the California Power Authority signed letters of intent for 315 MW of geothermal power at a price of 6 US cents per kWh.[50] As geothermal is used for baseload power, it can often command higher prices per kilowatt-hour than can wind or solar.

What's Next for Geothermal

Geothermal is a mature technology, which is both a blessing and a curse. It is proven and reliable yet unlikely to see major new technological improvements. Therefore, without policy change, it is unlikely to come in much lower than its current approximately 5 to 6 US cents per kWh. Also, its very limited resource will restrict development to just a few select countries.

Closing Thoughts

Renewable technologies, like fossil fuel technologies, all have their strengths and weaknesses. Many of the renewable technologies, however, are now realistic and practical alternatives to fossil fuel technologies. Costs are often higher for renewables than for fossil fuels, but their many advantages—reduced environmental impact, wider resource availability, sustainability—make policy change to promote renewables worthwhile.

Notes for Chapter 2

1. American Wind Energy Association (AWEA), "Wind Energy Outlook 2002," available at www.awea.org (downloaded 8 August 2002).

2. American Wind Energy Association (AWEA), "Wind Energy Outlook 2003," available at www.awea.org (downloaded 4 July 2003).

3. D.L. Elliott and M.N. Schwartz, "Wind Energy Potential in the United States," PNL-SA-23109 (September 1993), Pacific Northwest National Laboratories, 902 Battelle Boulevard, Richland, WA 99352, USA.

4. M. Grubb with R. Vigotti, *Renewable Energy Strategies for Europe, Volume II* (London: Earthscan Publications Ltd., 1997), p. II.84.

5. European Commission, Directorate-General for Energy, *Wind Energy—The Facts, Volume 1—Technology*, Executive Summary. Available from www.ewea.org (downloaded 25 January 2003).

6. M. Grubb with R. Vigotti, *Renewable Energy Strategies for Europe, Volume II*, p. II.85.

7. European Commission, Directorate-General for Energy, *Wind Energy—The Facts, Volume 1*.

8. "Great Expectations—Large Wind Turbines," *Renewable Energy World* (May-June 2001).

9. Based on a survey of turbine manufacturers in early 2001. See J. Beurskens and P. Jensen, "Economics of Wind Energy," *Renewable Energy World* (July-August 2001).

10. European Commission, Directorate-General for Energy, *Wind Energy—The Facts, Volume 2—Costs, Prices, and Values*, available from www.ewea.org (downloaded 25 January 2003), p. 7.

11. Assuming a turbine cost of US$675 per kilowatt.

12. Based on German and Danish experiences, and excluding turbines over ten years old. See J. Beurskens and P. Jensen, "Economics of Wind Energy."

13. P. Gipe, "Soaring to New Heights," *Renewable Energy World* (July-August 2002).

14. These were ten-year contracts. These contracts assumed that generators would receive the 1.7 US cents per kWh federal wind production tax credit, which was in place at that time. M. Bolinger and R. Wiser, Lawrence Berkeley National Laboratory, "Summary of Power Authority Letters of Intent for Renewable Energy," memorandum (30 October 2001).

15. "List of NFFO 5 Contracts for Wind Energy" (19 October 1998), from www.britishwindenergy.co.uk/ref/nffo5w.html (downloaded 7 November 2001). See also chapter 9 for more information on the UK's NFFO contracting process.

16. C. Archer and M. Jacobson, "Spatial and Temporal Distributions of U.S. Winds and Wind Power at 80 m Derived from Measurements," *Journal of Geophysical Research*, v. 108, no. D9 (2003).

17. Assuming module conversion efficiency of 10 percent, capacity factor of 22 percent, peak insolation of 1 kilowatt per square meter, and a total generating capacity of 800 gigawatts.

18. Some wastes may be produced in manufacturing and decommissioning.

19. Efficiency is defined as the percentage of energy in the sunlight striking the module that is converted into electricity.

20. J. Mortensen, "Factors Associated with Photovoltaic System Costs," NREL/TP.620.29649 (June 2001), available at www.nrel.gov (downloaded 21 June 2002), p. 2.

21. Solar Electric Power Association, "Large Systems Cost Report 2001 Update" (September 2001), available at www.solarelectricpower.org (downloaded 21 June 2002), p. 3.

22. R. Haas (editor), "Review Report on Promotion Strategies for Electricity from Renewable Energy Sources in EU Countries" (June 2001), available at www.itpower.co.uk (downloaded 20 January 2003).

23. M. Thomas, H. Post, and R. DeBlasio, "Photovoltaic Systems: An End-of-Millennium Review," *Progress in Photovoltaics: Research and Applications*, v. 7 (January 1999), p. 13.

24. U.S. Department of Energy and Electric Power Research Institute, "Renewable Energy Technology Characterizations" (December 1997), available at www.eren.doe.gov/power/techchar.html (downloaded 3 April 2000), p. 4.24.

25. This 40-US-cents-per-kWh result assumes the following: first cost of US$8,000 per kW, operations and maintenance cost of 1 US cent per kWh, lifetime of twenty years, cost of money at 7 percent per year, and a capacity factor of 22 percent.

26. U.S. Department of Energy, Energy Information Administration, "Renewable Energy 2000: Issues and Trends," DOE/EIA-0628(2000) (February 2001), available at www.eia.doe.gov (downloaded 21 June 2002), p. 31.

27. U.S. Department of Energy, Energy Information Administration, "Challenges of Electric Power Industry Restructuring for Fuel Suppliers," DOE/EIA-0623 (September 1998), available at www.eia.doe.gov (downloaded 13 November 2001), chapter 5, tables 14–17.

28. As measured by annual average total radiation on a horizontal surface. Data from J. Leckie et al., *More Other Homes and Garbage* (San Francisco: Sierra Club Books, 1981), p. 237.

29. U.S. Department of Energy, Energy Information Administration, "Renewable Energy 2000: Issues and Trends," p. 24.

30. Data are for 1999 and are from U.S. Department of Energy, Energy Information Administration. Data are available at www.eia.doe.gov.

31. Waste data from T. Johansson et al., *Renewable Energy: Sources for Fuels and Electricity* (Washington, D.C.: Island Press, 1992), p. 608. Actual electricity use from International Energy Agency, *Energy Statistics of OECD Countries 1997/1998* (Paris: OECD/EIA, 2000), pp. II.334, II.309, II.303.

32. "Electricity from Biomass, " Financial Times Management Report (1998), p. 35.

33. U.S. Department of Energy, "Economics of Biopower," from www.eren.doe.gov/biopower/basics/ba_econ.htm (downloaded 14 November 2002).

34. "List of NFFO 5 Contracts for Wind Energy" (19 October 1998), from www.britishwindenergy.co.uk/ref/nffo5w.html (downloaded 7 November 2001).

35. Per ton, over 100 years. See U.S. Congress, Office of Technology Assessment, "Changing by Degrees" (1991), available at www.wws.princeton.edu/~ota (downloaded 7 May 2002), p. 55.

36. U.S. Department of Energy and Electric Power Research Institute, "Renewable Energy Technology Characterizations" (December 1997), available at www.eren.doe.gov/power/techchar.html (downloaded 3 April 2000), p. 2-38.

37. U.S. Department of Energy and Electric Power Research Institute, "Renewable Energy Technology Characterizations," p. 2-24.

38. International Energy Agency, *Energy Statistics of OECD Countries 1997/1998*. Data are for 1998.

39. U.S. Environmental Protection Agency, "Landfill Methane Outreach Program" (21 August 2002), from www.epa.gov/landfill/projects/projects.htm (downloaded 12 September 2002).

40. P. Komor, "Renewable Technologies for Green Electricity Programs," E Source Green Energy Series GE-4 (May 2000), p. 26.

41. U.S. Department of Energy, Hydropower Program, cost data at http://hydropower.inel.gov/facts/costs-graphs.htm (downloaded 14 November 2002). Data inflated at 3 percent per year.

42. U.S. Department of Energy, Hydropower Program, cost data at http://hydropower.inel.gov/facts/costs-graphs.htm#prod (downloaded 14 November 2002).

43. U.S. Department of Energy, "U.S. Hydropower Resource Assessment Final Report," DOE/ID-10430.2 (December 1998), available at http://hydropower.inel.gov/state/stateres.htm (downloaded 15 September 2002).

44. Renewable Energy Plant Information System (REPIS) database, available at www.eren.doe.gov/repis (downloaded 5 April 2000).

45. G. Huttrer, "The Status of World Geothermal Power Generation 1995–2000," Proceedings, World Geothermal Congress (2000), p. 35.

46. C. Kutscher, "The Status and Future of Geothermal Electric Power," NREL/CP-550-28204 (August 2000), p. 2.

47. Author's estimate.

48. U.S. Department of Energy and Electric Power Research Institute, "Renewable Energy Technology Characterizations" (December 1997), chapter 3.

49. Idaho National Engineering Laboratory, Geothermal Energy Web site, "How Much Does Geothermal Energy Cost per kWh?" available at http://id.inel.gov/geothermal/faq/q02.html (downloaded 14 November 2002).

50. Based on data collected by Ryan Wiser at Lawrence Berkeley National Laboratory.

3

Electricity Restructuring and Renewables: A Primer

Until the late 1980s, electricity systems in the industrialized world were either government-run or heavily regulated, with new generation investments and retail electricity prices set or overseen by government rather than by market forces. Since then, however, many industrialized countries have restructured ("liberalized") their electricity systems, injecting competition and market forces into the generation and retail parts of the system. The European Union (EU) has a directive (requirement) that member states open up their electricity systems to competition, and most EU member countries are doing so. Many U.S. states have introduced some form of competition into their electricity systems, including California, Pennsylvania, and Texas. This transition is a difficult one, with some areas—notably California—suffering from high electricity prices and shortages.

This wrenching change is both an opportunity and a challenge for renewable electricity—as it opens many potential paths for renewables to play a larger role in electricity systems and closes some paths as well.

This chapter explains how electricity systems work, notes the factors that are driving the ongoing change from regulated to competitive systems, and discusses how countries' electricity systems are evolving in response to these drivers. It closes with a discussion of restructuring's implications for renewable electricity.

A Brief Note on Terminology

Is a "supplier" of electricity also a "retailer"? How do "domestic" energy users differ from "residential" energy users? What is "harmonization"? Continental Europe, the UK, and the U.S. often use different terms that, in fact, mean the same thing, which adds to the confusion and makes cross-country comparisons all the more difficult. Here we sort out a few of the most confusing terms.

Renewables are electricity-generating technologies that use renewable resources (wind, sunlight, water, and biomass) to make electricity. Renewables are similar but not identical to green electricity technologies.

Green is a fuzzy word, meaning different things to different people. It implies a reduced environmental impact. In the case of electricity-generating technologies, most—but not all—renewable electricity-generating technologies are generally considered green. Large hydro is the exception: It is undeniably renewable, but many do not consider it green.

End users are the actual purchasers of electricity—those who use electricity to make heat, light, and other useful services. Synonyms are *consumers* and *customers*.

Providers are companies that provide or sell electricity to end users. Some texts use the terms *suppliers* or *retailers*. However, *suppliers* is also sometimes used to mean companies that generate electricity. This book avoids using the word *suppliers* entirely, given its great potential for causing confusion.

Restructuring means the entire process of shifting electricity systems from heavily regulated, monopolistic structures to ones involving competition in generation, free market entry for new generators, and choice of providers for end users. Synonyms are *liberalization, deregulation,* and *competitive transition.*

Residential (as in "the residential sector") refers to electricity end users purchasing electricity for home use—for lighting, cooking, heating, and so on. A synonym is *domestic.*

Noncompetitive electricity systems are those in which government agencies or regulated monopolies provide electricity.

Harmonization means the broad policy goal of having similar policies across EU member countries.

From Fuel to End Use: An Overview of Electricity Systems

Electricity systems start with the *fuel,* such as coal, water (for hydropower), or uranium (for nuclear power). This fuel is delivered to the *power plant.* The power plant converts the fuel into electricity. This conversion of fuel into electricity is called *generation.* The electricity is then sent into the *transmission* system, which moves electricity from the generating facility to the area where it's needed—from the coal-burning power plant to the city, for example. The *distribution* system carries the electricity from the end of the transmission system to the end user—a house, an office building, or an industrial facility. Once it gets to the end user, electricity is typically converted again, this time into something useful—light, heat, motor drive, or other *end uses.* (See **Figure 3-1.**)

Figure 3-1: From fuel to end use: Components of electricity systems

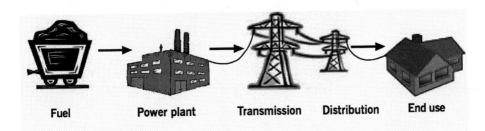

| Fuel | Power plant | Transmission | Distribution | End use |

The generation part of the system is responsible for most of the electricity-related environmental damage, most of the cost of electricity, and most of the controversy involving electricity. It is also the part in which renewables could play a major role.

Before the Revolution: Noncompetitive Electricity Systems

Until about 1990, most electricity systems in industrialized countries were either run as public or quasi-public agencies (as in the UK and France) or as tightly regulated, privately owned companies (as in the U.S.). Although this is changing, understanding this change requires first looking at how it used to be, and in some areas still is, and figuring out how and why the old noncompetitive systems worked.

Electricity was long considered a "natural monopoly"—an industry for which it made sense to have just one firm doing the work. It wouldn't make sense to have multiple wires running to each house, so the argument went, so just one firm should own and operate the electricity system. There were other arguments as well in favor of a monopolistic system: The provision of dependable, low-cost electricity was seen as both a public right and a requirement too important to leave to the vagaries of a competitive market; the failure of a national electrical system would have catastrophic economic and other effects, and therefore government should control or regulate the system; and the capital requirements for some system components, such as large power plants or new transmission lines, would require government underwriting. There was also a historical public mission in some areas of extending electricity service to those regions without electricity—an action that was clearly of societal benefit but that required some subsidy as well. Although these arguments may seem out of fashion today, it's interesting to note that they are coming back as problems emerge with the new competitive system (see for example the box on California further on in this chapter).

The actual electricity industry structure varied by country. In the UK, one organization, the Central Electricity Generating Board (CEGB) controlled and operated all generation and transmission. Twelve regional electricity boards handled distribution.[1] The relationships among the CEGB, the distribution boards, and the government itself were complex,[2] but the CEGB was clearly a nationalized industry, with a mission to serve the public rather than to make a profit. In the Netherlands, generation was dominated by four companies, which were largely owned by local and regional governments. In France, one large organization, Electricité de France, handled almost all aspects of electricity. Other EU countries used similar nationalized or tightly regulated industry structures.

In the U.S., most electricity was provided by what were called "electric utilities," which were investor-owned companies responsible for generation, transmission, and distribution. These companies were vertically integrated, meaning that they were responsible for electricity all the way from its generation to its delivery to the end user. State (not federal) governments regulated these companies. The main lever of regulatory control was via the prices these utilities were allowed to charge their customers, but other decisions, such as those concerning power plant investments, went through the regulators as well.

This noncompetitive system generally promoted slow, methodical, and relatively open decision making. Firms were discouraged from taking financial or technical risks, being innovative, or offering new products or services. The emphasis was on dependability and minimizing (perceived) risk. There was a "regulatory bargain" in the U.S.: The utility provided dependable and reasonably priced electricity and was, in turn, allowed to charge prices for its electricity that allowed it to provide a reasonable return to its investors.

These firms were often used to promote or pursue public or societal goals as well. The UK's CEGB was at times pressured to maintain domestic coal industry employment by paying higher-than-market prices for domestic coal.[3] In the U.S., utilities in some areas were directed by regulators to invest in energy efficiency and renewable energy, although these investments were quite small relative to the utilities' traditional supply-side spending.

In most countries, the old noncompetitive system generally succeeded at yielding dependable, reasonable-cost electricity for users. There were exceptions, but most such noncompetitive systems experienced relatively few outages, and electricity prices were stable. Some argue that electricity prices would have been lower if the system had been subject to competitive price pressures, but this is uncertain.

The old system's risk aversion showed in its lack of technical innovation. Companies under regulation were very hesitant to adopt new technologies. An example of this is the still-common use in the U.S. of analog manual-read electricity meters. New remote-read meter technologies are cheaper, more accurate, and more flexible (allowing for remote analysis of consumption patterns), but the regulated utility had little reason to adopt new technologies such as digital meters. There was no competitive pressure, no call from customers for technologies they were not even aware of, no interest from regulators in increasing risk, and no reward to the utility for doing anything markedly different from what it had been doing for years.

Because of the regulatory bargain, capital was relatively inexpensive: Utilities were essentially guaranteed their money back by the regulators, so lenders saw little risk in lending to utilities. It was therefore possible for utilities to build very large power plants with long construction times. This practice was also thought to be less expensive overall due to economies of scale.

Forces for Change

The old noncompetitive system described above was certainly not a failure; it provided dependable, reliable, and reasonably priced electricity. However, starting in the early 1980s and accelerating through the 1990s, several factors combined to cause a fracture and, in some cases, complete disintegration of this system and a rise of a very different system based on competitive markets with multiple electricity producers and a smaller government role.

One of the first cracks in the system appeared in 1978 in the U.S., when the federal government passed a law requiring utilities to purchase electricity from other companies. Prior to that law, it was essentially impossible in the U.S. for a company other than a utility to go into the electricity generation business. The utility controlled the entire electricity delivery system, and short of running wires to the end user, there was no way for a non-utility generator to get into the system. Utilities had no interest in buying electricity produced by another company, as they had their own electricity-generating plants. Potential new generating companies argued, however, that utilities should buy their electricity as long as it was less expensive than the electricity that utilities produced themselves—an appealing, if simplistic, argument. So the U.S. federal government passed a law (called PURPA, the Public Utility Regulatory Policies Act) that required utilities to buy electricity from non-utility generators and, furthermore, to pay them the "avoided cost"—meaning, in theory, what it would have cost the utility to generate that electricity itself. (More information on PURPA can be found in Chapter 8.)

At about the same time, several technological advances also began to exert an influence. The most important was the availability of small, efficient, and low–capital cost natural gas–fired turbines. All of a sudden, it was possible to buy an industrial-sized natural gas–fired power plant and, depending on prevailing natural gas prices, generate electricity for less than that charged by the utility. This made the centralized, regulated, monopoly utility structure look outdated and expensive. Improvements in cogeneration technologies (which simultaneously produce electricity and useful heat) also increased political pressure on the old systems by making it increasingly attractive for larger industrial users to generate their own electricity.

Problems with nuclear power contributed to the change as well. The nuclear story is a long and complex one, and the finger-pointing continues to this day. But the essence of the story is that nuclear power in its early days was seen as a safe, inexpensive, dependable, and clean technology. As a result, many countries invested heavily in it. The reality, unfortunately, turned out to be quite different from the expectations. For a variety of reasons, in most countries nuclear power ended up being expensive, unpopular, and, in some cases, undependable as well. There is considerable disagreement over why the economic and political realities did not match the apparent technical potential. Regardless of what one might believe about who is at fault and who should be held responsible, however, the nuclear experience left many with the view that the government-dominated electricity system made poor financial and technology decisions.

The existence of glaring price differences across geographic borders also contributed to political pressure for change. Some U.S. states with nuclear power had much higher electricity prices than neighboring states without nuclear generation. This led to industrial users threatening to move out of the expensive states, which in turn led to political pressure for price reduction and regulatory change.

Perhaps the most important driver was the global trend toward privatization. In the 1980s, many industrialized countries embarked on a mission to privatize many of their energy, transport, and telecommunications-related public services. In the UK, for example, the long list of formerly public or monopolized industries that have been privatized includes British Aerospace (1981), British Telecommunications (1984), British Gas (1986), British Airways (1987), British Steel (1988), British Coal (1995), and British Rail (1996). The causes of this privatization were wide and varied, from unions' loss of political power to shifts in beliefs about the appropriate role of government to users' dissatisfaction with the quality of services provided. The global trend to privatization, in any case, left few industries untouched, and electricity was one of its primary targets.[4]

Response: Introduction of Competition

The factors described above drove governments throughout the industrialized world to open up their electricity systems to competition. This section describes three distinct approaches to restructuring: (1) the EU approach, which sets an overall schedule for member countries; (2) the U.S. plan, which is progressing in fits and starts; and (3) the UK model, which opened the electricity market early and all at once.

Restructuring in the EU

As one European research report noted, "There is little controversy about the gradual opening of the [electricity] market. The idea is that opening of the market will lead to more competition, better services and customer orientation and cost efficiency. Price differences that are not based on cost differences will have to disappear."[5] The passing of EU directive 96/92/EU in December 1996 evidences the EU-wide agreement with this view. This directive, which is essentially a law that all EU member countries are compelled to observe, requires member countries to open up their electricity markets to competition. The directive sets up a schedule that all EU member countries must meet. Although different countries are following quite different paths to an open market, most are making rapid progress and are ahead of the EU directive's requirements (see **Table 3-1**).[6] In the Netherlands, for example, the Electricity Act of 1998 directed that large industrial users with a demand of 2 megawatts (MW) or greater be the first to be able to choose their electricity supplier, followed by medium users in 2002, and finally all users by 2004.[7] Since the passage of that act, the various generation and distribution companies in the Netherlands have been through a whirlwind of mergers, buyouts, takeovers, and related corporate positioning.

Table 3-1: Selected EU countries' restructuring status

Country	Date of full retail choice
Belgium	2007
Denmark	2003
Ireland	2005
The Netherlands	2004
Spain	2003
UK	1998

Restructuring in the UK

Up until 1990, the electricity system in England and Wales was publicly owned. The Central Electricity Generating Board (CEGB) operated all the generation and transmission facilities, while distribution was handled by twelve Area Boards, each operating as a monopoly in its service area. This cozy system was deemed inappropriate by the government of Margaret Thatcher, however, which had a strong bent for privatization. The government's efforts resulted in the Electricity Act of 1989, which divided the UK's electricity system into parts for subsequent sale.

The generating side of CEGB was split into three companies (PowerGen, National Power, and Nuclear Electric), which were to be sold to the public. PowerGen and National Power were sold off in part in 1991, and the remainder sold in 1995. Nuclear Electric, however, proved problematic, as its very high generation costs—estimated at two to three times that of fossil-fired plants—and its uncertain future liabilities of decommissioning and waste management attracted a distinct lack of willing buyers. After considerable uproar, Nuclear Electric received a public subsidy of about US$14 billion, collected via the Fossil Fuel Levy.[8] The twelve Area Boards that handled electricity distribution were sold off in 1990. The transmission side of CEGB was transferred to the recently privatized Area Boards.

Restructuring implies competition, as well as privatization, and the UK has seen retail competition in electricity as well. Very large customers (over 1 MW) were free to choose their suppliers starting in the early 1990s. Competition for remaining customers was phased in over the next few years: medium-size users (100 kilowatts to 1 MW) were given choice in 1994, and domestic and small business customers were given choice in a phased process from 1998 to 1999.

One useful measure of the success of restructuring is the number of end users who switch providers. Switching in the UK has been robust, suggesting that there is true competition in electricity retailing. By 2000, over half of commercial and industrial electricity users had switched providers.[9] The domestic (residential) market did not open up until 1998–1999, but saw relatively rapid switching as well. By late 2001, over one-third of residential consumers had switched electricity providers.[10]

Restructuring in the U.S.

The U.S. is following a more meandering path to a competitive and open electricity system. This is due in part to the fact that the bulk of U.S. electricity regulation is at the state rather than national (federal) level.

There have been numerous attempts to pass legislation in the U.S. Congress that would, in effect, force states to restructure. Most such legislation, if passed, would function like the above-mentioned EU Directive: It would set a schedule by which states would have to open up their electricity markets. However, such legislation has not come close to passing, and given California's 2000 restructuring disaster (see box below), it is not likely to. Therefore, the 50 states have been left to make their own decisions about restructuring.

The drivers listed above (notably new technologies and overall trends for privatization) all played a role in motivating U.S. states to restructure, but perhaps the most important was the perception of high and unfair electricity prices, especially in some areas where neighboring states had much lower prices. This resulted in considerable political and economic pressure on some state governments to take action to reduce the threat of industries moving out of state in search of lower electricity prices.

The first U.S. state to open up its electricity system was California, in which consumers were given choice in March 1998. (Choice was rescinded in 2001, see box below.) Montana, Pennsylvania, New York, and a number of other states soon followed. In the late 1990s, it appeared that most states were either restructuring or had firm plans to do so. However, California's experience put a halt on many states' restructuring plans. As of early 2003, seventeen states had introduced competition or had set explicit schedules for doing so. An additional six had delayed or cancelled their restructuring plans. The remaining twenty-seven states had no explicit plans for restructuring (**Figure 3-2**).[11]

Figure 3-2: State-by-state status of restructuring in the U.S. as of February 2003

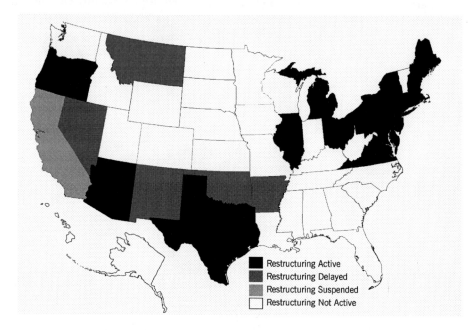

Legend:
- Restructuring Active
- Restructuring Delayed
- Restructuring Suspended
- Restructuring Not Active

California's Electricity Restructuring Debacle: What Happened?

California was the first U.S. state to open up its electricity system to competition. In 1996, the California state government passed legislation that, among other changes, allowed electricity end users to choose their electricity providers beginning in March 1998. The new system worked fairly well until the summer of 2000, when it failed miserably. At that time, the system disintegrated, with dramatic effects:

- Wholesale electricity prices increased radically—by a factor of ten or more.[12]

- Power shortages struck, requiring blackouts.

- The three major utilities in the state experienced very severe financial crises, and one—Pacific Gas and Electric—filed for bankruptcy.

What caused this catastrophe? There were a number of factors, any one of which the system probably could have weathered, but all at once proved too much. Among the major contributors to the crisis were the following:

- No new generating capacity was built in California in the 1990s, although electricity demand grew markedly during that time.

- Low rainfall levels led to about a 3-gigawatt (GW) reduction in hydropower availability.[13]

- Approximately 10 GW of California's generating capacity was not operating at peak times.[14]

- There were transmission capacity constraints between northern and southern California.

- When the three major utilities began to suffer financial problems, independent generators were hesitant to sell power to them as they feared not getting paid.

- Natural gas prices were high.

- Utilities were not allowed to sign long-term contracts and were therefore forced to buy high-priced power on the spot market.

- The utilities were not allowed to pass on these costs to their customers.

The state's electricity crisis triggered a political crisis as well, as the state government struggled to stabilize the system. A number of actions were taken, including changes in wholesale buying and selling rules, rate increases, streamlining of environmental regulations, and the establishment of a new state agency to buy power. The most dramatic step was the decision in September 2001 to essentially take away choice: As of that date, end users could no longer choose their providers. Noted one state regulator, "It is essentially hollow to claim that the delivery of (electricity) service, which is essential to the functioning of society, should be turned over to the whims of the free market....It hasn't worked."[15] Thus, the earliest and most ambitious restructuring program in the U.S. ended, ironically, with an electricity system with a much *greater* public role than the one it initially replaced.

What Does Restructuring Mean for Renewables?

What does this fundamental shift in electricity systems—from regulated monopoly and direct government ownership to competitive, open markets—mean for renewables? It creates a plethora of new opportunities, but it closes off a few cherished ones as well.

New Opportunities

In the past few years, as electricity systems have begun to open up, renewables have been booming. This is not a coincidence—the new competitive electricity market has been very good for renewables. These new opportunities include:

- the ability to differentiate retail products such as green power;
- incentives for providers to seek competitive advantage, through, for example, branding;
- new policy tools such as renewable obligations and green certificates;
- greater preference for new generation technologies with shorter construction times; and
- the potential for the removal of political borders to electricity, thereby reducing the impacts of unequal renewable resource distribution.

Product differentiation and branding. A competitive electricity system allows for product differentiation, which means consumers can be offered a variety of products and services rather than just one, undifferentiated product from a single monopoly provider. Although this may sound irrelevant to the commodity-like product of electricity, there are already a number of differentiated electricity products: "green" (renewable) electricity, "high-quality" (stable voltage, low electrical noise, very high reliability) electricity, and others. The demand for green electricity is surprisingly robust, and this is in turn driving construction of new renewable-based electricity generation.

Similarly, differentiated products can help competitive providers brand themselves in a competitive electricity system. Although competitive electricity markets are still young, some electricity providers have decided that the way to profits lies not in being the lowest-cost provider of commodity electricity but in providing higher-value electricity-related products and services. For some providers, these include green electricity, home renewable systems, and other renewables-related offerings. Similarly, some providers are supporting renewables in an effort to brand themselves as the environmentally friendly electricity company.

Policy. The transition to a competitive electricity system has stimulated the creation of a number of imaginative policies and approaches to encouraging renewables. These policies, notably renewable obligations (also called renewable portfolio standards, see chapter 10) are already resulting in significant new renewable electricity generating capacity. These policies, along with others directed at energy efficiency, low-income electricity users, and other public benefits are interesting in that they represent more, rather than less, regulation and government role in the electricity industry. Thus, restructuring is a better descriptor than "deregulation," as regulations change significantly, but they certainly don't go away.

Shifting investment criteria. In today's more competitive system, new generation capacity investments are increasingly made by competitive companies looking to make a profit. These firms will, all else being equal, prefer investments that pay off sooner rather than later. Unlike the old regulated utilities, they are more capital-constrained and unable to convince capital markets of the wisdom of building very large new plants that take many years to complete. So the investing preference is shifting to smaller generation plants with shorter lead times. Renewable power plants (other than large hydropower) are, in most cases, small and relatively quick to build. They are also modular, meaning that an initial block of renewable capacity can be up and running relatively quickly, with more capacity then added as appropriate.

Part of the logic behind the competitive transition, especially in the EU, is the removal of geographic barriers to electricity markets. Although a moderate amount of electricity is already traded internationally in the EU (the Netherlands, for example, imports 16 percent of its electricity, mostly from Germany and France[16]), international trade likely will increase as electricity markets liberalize. This could benefit renewables by reducing the effects of unequal distribution of renewable resources. For example, the output from offshore wind turbines in the North Sea could be sold to inland countries that lack wind resources.

New Challenges

There are a number of ways that restructuring will make life *more* difficult for renewables:

- Reduced regulatory control will make it harder for governments to use utilities as a tool to promote social goals such as increased use of renewables.

- The new competitive power market has a shorter time horizon and higher discount rate for investments, both of which, all else being equal, reduce the impacts of future fuel costs and thus favor fossil-fueled technologies over renewables.

- Similarly, smaller competitive companies will be more capital-constrained than regulated utilities and thus will prefer lower–capital cost technologies.

- As the wholesale electricity market becomes more liquid and transparent, the intermittent nature of some renewables will become more visible and may reduce the perceived value of these renewables. Similarly, increased use of bilateral contracting may result in reduced prices for intermittent technologies.

Reduced regulatory control. In the old noncompetitive system, it was quite easy for the government to carry out societal goals related to renewables. Consider, for example, the situation in which the government decides that, because of environmental externalities,[17] it is appropriate to have a wind turbine as part of the country's electricity generation mix. In the case of government-managed or government-run utilities, the utility simply installs the turbine; for regulated monopolies, as in the U.S., the regulator directs the utilities to install the turbine and allows the costs to be charged back to the ratepayers. In a competitive system, however, the government has no such direct control over electricity generators. It can (and does) try to influence their behavior, of course, via various policies, but its influence is less direct and less certain than under the former system of direct control.

Discount rates and capital constraints. As noted above, electricity market liberalization means investors will look to smaller generation technologies with shorter lead times. Investors will also use higher discount rates and be more capital constrained than government or regulated utility investors. Both of these factors can work *against* renewables. As discussed in Chapter 2, most renewables have higher initial (capital) costs but lower operating (notably fuel) costs than fossil-fueled generation. Higher discount rates, all else being equal, shifts preferences toward choices with lower capital costs and higher future (operating) costs, which might mean the selection of gas turbines over wind turbines. Similarly, small investors and new market entrants may be capital constrained and therefore looking to put in the lowest–capital cost option, which is unlikely to be renewables.

Intermittency. An increasingly important issue is the intermittence, or lack of dispatchability, of many (not all) renewables. This has always been a potential problem, but in the old regulated system it was less of a concern, because intermittent renewables (wind and photovoltaics) played a minor role in the overall generation mix, and vertically integrated utilities could easily use other generation to smooth out the total system output. Some restructured electricity systems, in contrast, encourage direct bilateral contracts between generators and consumers. As is already becoming apparent in the UK, however, bilateral contracting can sharply reduce the perceived value of intermittent renewables.

Summary

Restructuring is moving along briskly in the EU, and although it stumbled mightily in the U.S. due to California's problems, it is limping along there as well. Restructuring is neither the death of nor the savior for renewables, it is instead a new environment in which renewables must compete. Renewables are doing well in already-restructured regions, but continuing progress will require jumping on the opportunities restructuring creates, while letting go of the old policy levers as they fade away.

Notes for Chapter 3

1. This discussion applies to the electricity system serving England and Wales. Scotland and Northern Ireland had similar but separate systems.

2. For a nicely detailed discussion of this relationship, see J. Chesshire, "UK Electricity Supply Under Public Ownership," in J. Surrey, ed., *The British Electricity Experiment* (London: Earthscan, 1996).

3. See J. Chesshire, "UK Electricity Supply," p. 32. Also U.S. Department of Energy, Energy Information Administration, "Electricity Reform Abroad and U.S. Investment," available at www.eia.doe.gov/emeu/pgem/electric/ch2.html (downloaded 10 February 2003), chapter 2, p. 2.

4. Note, however, that privatization is not inevitable, universally popular, or always a success. California's electricity restructuring problems led to reregulation, while problems with the UK's privatized rail system led to calls for its renationalization.

5. ECN (The Netherlands Energy Research Foundation), "Energy Market Trends in the Netherlands—2000" (June 2000), p. 13. See www.ecn.nl.

6. International Energy Agency, "Energy Policies of IEA Countries Compendium 2002" (December 2002), table 9. See www.iea.org.

7. ECN, "Energy Market Trends in the Netherlands—2000," p. 34.

8. G. Mackerron, "What Can We Learn from the British Nuclear Power Experience?" in G. Mackerron and P. Pearson, *The UK Energy Experience: A Model or a Warning?* (Imperial College Press, 1996), pp. 247–258.

9. Ofgem (Office of Gas and Electricity Markets), "A Review of the Development of Competition in Industrial and Commercial Electricity Supply Again" (December 2000), pp. 13–15.

10. Ofgem (Office of Gas and Electricity Markets), "Competition Explained Main Page" (12 November 2002), from www.ofgem.co.uk (downloaded 25 February 2003).

11. U.S. Department of Energy, Energy Information Administration, "Status of State Electric Utility Restructuring Activity" (February 2003), available at

www.eia.doe.gov/cneaf/electricity/chg_str/regmap.html (downloaded 3 July 2003).

12. December 2000 wholesale prices on the power exchange were $3,780 per MWh, over eleven times higher than the December 1999 price of $30 per MWh. U.S. Department of Energy, Energy Information Administration, "California Electric Energy Crisis" (8 August 2002), from www.eia.doe.gov/cneaf/electricity/california/subsequentevents.html (downloaded 10 February 2003).

13. U.S. Department of Energy, Energy Information Administration, "California's Electricity Situation," Briefing for the Staff of the U.S. House of Representatives (9 February 2001), from www.eia.doe.gov/cneaf/electricity/california/california.html (downloaded 10 February 2003).

14. This was due in part to collusion by generators attempting to drive prices up.

15. As quoted in M. Bazeley, "Regulators Eliminate User Ability to Pick," in *San Jose Mercury News* (21 September 2001).

16. ECN, "Energy Market Trends in the Netherlands—2000," p. 13.

17. Environmental externalities are those environmental impacts that are not reflected in the market price. For example, a coal-burning power plant emits SO_x, which causes some environmental damage, but the costs of this damage are generally not reflected in the price of the electricity produced by the plant.

4

The UK Green Electricity Market: Why Was It a Flop?

Green electricity programs and policies—in which end users are given the option of purchasing renewable-sourced electricity, typically at a 1- to 3-US-cents-per-kWh premium—are increasingly popular. We take a close look at such programs in three countries: the UK (this chapter), the U.S. (chapter 5), and the Netherlands (chapter 6), and in addition examine how and why individuals and organizations choose to pay more for their electricity (chapter 7).

Green electricity products in the UK have not been popular. Although green electricity has been available in most of the UK since the advent of retail choice, as of August 2002, only 60,000 residents had signed up for it—about 0.2 percent of all UK households.[1] This was far behind the sign-up levels in the Netherlands (900,000) and Germany (250,000) at that time. This is especially surprising in light of the many factors that suggest the UK's green market ought to be a roaring success:

- The UK has a robustly competitive electricity market, with over one-third of all residential users having switched providers as of late 2001.[2]

- Several new green providers entered the UK market, including some innovative, market-savvy, and assertive new companies.

- The UK has a reasonably aggressive national renewables goal, resulting in notable policies to promote renewables.

- Renewable resources in the UK are sufficient to supply the UK's electricity several times over.

The above list sounds like a recipe for success rather than one for failure—making it all the more worthwhile to take a closer look at what happened to the UK green energy market. As is shown in this chapter, making a green market work is a tricky business. Although the items on the above list are certainly useful in building a green market, they are unfortunately not sufficient to ensure green success.

Description of the UK Green Market

As discussed in chapter 3, the UK opened its electricity system to retail choice in phases—starting with large industrial users in the early 1990s, and ending with residential users in 1998–1999. Switching has been robust, with over half of all commercial and industrial users and one-third of all residential users switching as of late 2001.

The availability of green electricity products has tracked the availability of choice. The UK's first green product was introduced in 1996. In that year, the Renewable Energy Company offered "Ecotricity"—a green product of mostly landfill gas–sourced electricity—to commercial electricity users. (At that time, residential users did not yet have choice.) As choice grew to encompass the residential market, the incumbent electricity suppliers began to offer green products as well. By 2000, all the original incumbent suppliers as well as a handful of new market entrants offered green electricity. Green products change often as companies refine and modify their green offerings, but by 2000, almost all UK electricity users had the option of purchasing green electricity.

The UK green offerings are of two main types: "source" and "fund." Source programs are the usual form of green electricity. In such a program, the provider commits to wholesale purchases of renewable electricity equal to that purchased by their customers. For example, London Electricity offered a "green tariff" to their business customers for a 0.4 pence (0.6 US cents) per kilowatt-hour (kWh) premium over their regular system electricity.[3] Fund programs are those in which extra money is collected from customers to support a fund used to build new renewable capacity. For example, Eastern Energy offered customers the option to contribute an additional 50 pence (75 US cents) per week toward a fund that is used to support new renewable construction.[4] Of the sixteen green products offered in the UK in 2000, eight were source, five were fund, and three were combination products.[5]

Most of the UK's green products were offered by default providers—that is, by electricity providers who were formerly the monopoly providers. These default providers offered green as an alternative to their regular, brown electricity. Although their green products varied in terms of fuels used, target audience, and pricing, they often had the common characteristic of lackluster marketing (see discussion later in this chapter).

The new entrants to the market had products that were a bit more varied. As of 2000, for example, the Renewable Energy Company was selling only green electricity and only to commercial customers, although it was considering the domestic market as well. It was selling 100 percent green but was considering a mixed product (partly green, partly fossil based) in the future. At that time, an increasing number of UK commercial-sector electricity users were using a tender[6] to buy electricity. Although the Renewable Energy Company did not claim to be the lowest bidder, it considered itself a competitive bidder with the added attraction of offering green electricity.[7]

Unit[e], another new market entrant, also sold only green electricity. It sold small hydropower (10 megawatts [MW] or less) and wind-based electricity at a premium that was about 12 percent above regular brown rates, but that premium was expected to come down.[8] Unit[e]'s focus on environmentally friendly renewable technologies led to its ranking as the "greenest" UK offering by the environmental group Friends of the Earth.[9]

Reason 1 for Flop: Focus on Cost

The robust switching market in the UK provided useful insights into what effect individual behavior has on electricity markets and green power. Analyses of data from this market reveal who is switching, why, what information they use to select a supplier, and how green fits into their decision process. In this case, analysis of consumer switching data reveals one of the key reasons why the UK green power market got off to such a slow start. In short, switching in the UK was presented and defined as a way to save money; whereas, buying green is something quite different.

Initially, switching rates were proportional to income—that is, those with higher incomes were more likely to have switched suppliers.[10] However, after two years of choice, that pattern faded and switching rates became similar across income groups, with the exception of very low income households, which still lagged the rest of the population.[11] Indicators that still differed for switchers and nonswitchers include those on prepayment meters and those without bank accounts, both of which correlated with lower switching rates (and lower income as well).[12]

So why were some people switching and others not? Interviews with non-switchers revealed that "the principal reason for not switching electricity supplier is inertia...many nonswitchers are satisfied with their present supplier and see no need to change."[13] The interviews suggested that nonswitchers were, in general, aware that they had a choice and not intimidated by the switching process. They just didn't think it was worth it.

The most striking finding from consumer interviews was the preponderance of both *price* and *cost* as drivers for switching. Price was the key variable consumers used to distinguish among potential suppliers; price was the reason most consumers decided to switch; and price will drive future switching decisions as well:

- "Seventy percent of switchers spontaneously say that they changed supplier to take advantage of cheaper prices."[14]

- "Among those who have switched...the single biggest motivator for switching suppliers is price. Indeed, this factor has been top throughout the previous years' research."

- More than two-thirds of nonswitchers say cheaper bills would be the most important factor in their decision to switch suppliers.[16]

Consumers' focus on price reflects the political context of restructuring in the UK. Restructuring, and indeed privatization in general, was sold on the argument that it would increase economic efficiency and reduce costs to consumers, while maintaining or even improving service. Although the political winds shift (for example, difficulties with the UK's privatized rail service disrupted plans to privatize London's Underground subway system), the focus throughout the UK's privatization efforts was the potential for price reduction. This was the focus of early press coverage and the focus of marketing by competing suppliers. In addition, government efforts to educate and inform consumers about choice portrayed price as *the* criteria for choosing a supplier.

Why is this significant? Before restructuring, of course, there was no choice. Then, when restructuring began to enter the public arena, it was presented as a way to save money. Initial impressions are critical, and consumer understanding of restructuring was built upon this base. Since then, the message that restructuring is all about saving money has been reinforced by the government, by the press, and by the retailers themselves.

One could imagine, in contrast, a different but equally plausible scenario in which restructuring is seen by consumers as providing choice across many variables: dependability of service, billing options, greenness of supply, power quality, price, and so on. In such a scenario, retailers would compete across the variables.

As it is, however, green retailers face a daunting task: They have to first convince consumers that their understanding of choice only as a way to save money is not accurate, and they have to then sell their product on the new variable of greenness. Given this, it's no surprise that green products have not done well in the UK.

There are other reasons why the UK green market has foundered, as discussed below, but an important one is that choice in the UK was initially presented, and has continued to be seen, as a way to save money. Breaking consumers out of this mindset will remain a challenge.

Reason 2 for Flop: Lousy Marketing by Incumbents

Interviews with the default suppliers about their green marketing efforts found that suppliers "considered the market too small to warrant an extensive marketing campaign,"[17] which is, of course, a self-fulfilling prediction because marketing is needed to grow the market. Examples of their meager marketing efforts include the following:[18]

- "The marketing of Company G's green electricity product has been limited to a feature in their consumer loyalty magazine."

- "At Company H, marketing is limited to sending out brochures to customers who specifically ask about GE [green energy]."

- For Company M, "marketing for commercial customers has not seemed necessary because 'commercial customers are asking for it of their own accord.'"

Experience in the U.S. has shown that sales levels for green electricity (and for consumer products generally) are a direct function of marketing effort (see chapter 5). So why were the default providers in the UK not making a serious effort to sell green energy? There were several reasons. First, these companies were used to serving a monopoly market and had no need for or interest in marketing. The transition from a monopoly market with guaranteed customers to a competitive market is a difficult one. In other words, these companies did not have enough of a marketing mindset to recognize that, if you want customers to buy something, you have to sell it to them. Second, some of these companies saw green energy as a narrow and limited niche market and believed that the potential market was too small to make it worth investing in a serious marketing campaign. Third, some expressed the concern that the supply of green electricity was too small to support a larger green market—that if they did manage to sign up more customers for green energy, they would be unable to deliver it.[19] (The U.S. experience has shown a straightforward fix for this—sign customers up but don't charge them, and use this "presold" capacity as a lure to get new capacity built quickly.)

Not all UK green marketing was lackluster. The green product "RSPB Energy" was quite successful and is a nice example of shrewd marketing. RSPB is the Royal Society for the Preservation of Birds, Europe's largest conservation charity, with more than one million members and 150 nature reserves.[20] RSPB had a relationship with the electricity provider Scottish and Southern Energy plc to co-brand a green electricity product. RSPB Energy offered 100 percent renewable-sourced electricity, mostly from hydropower. In addition, the RSPB received US$14 initially and US$7 per year for each customer signing up for electricity service. RSPB marketed the program several ways, including advertising in its quarterly members' magazine. It had about 14,000 domestic customers buying electricity through the program in mid-2002, making it one of the UK's most successful green electricity programs at that time.[21]

Reason 3 for Flop: Policy Conflicts and Uncertainty

In the early 2000s, the UK's renewable energy market was in the midst of policy chaos. Three significant new renewables policies—the Climate Change Levy, the New Electricity Trading Arrangements, and the Renewables Obligation—were recently underway or planned for the near future. Although these policies were intended to drive the market toward more renewable energy in the long term, their short-term effects on the voluntary green market were chilling. The interplay between these various policies nicely illustrates the complexities and unintended side effects of policy change.

These three pro-renewables policies ended up causing problems for the voluntary green market primarily by increasing uncertainty, which led to increased perceived risk, which led to reduced investment in marketing and reduced overall market activity. Many green marketing efforts were put on short-term hold pending greater clarity on just how the new policies would fall out. A leading green marketer, for example, reported that its corporate parent, which provides its working capital, had decided to hold off on funding ambitious marketing plans until the policy details were made clear.[22]

The Climate Change Levy

The climate change levy (CCL) was a simple policy with complex effects. It was a tax on energy use, with the goal of increasing energy efficiency and helping to meet the UK's carbon reduction targets. Starting 1 April 2000, electricity users in the UK began paying an additional tax of 0.43 pence per kWh (0.62 US cents per kWh) on their electricity usage. The CCL is not a trivial tax: It works out to be 11 percent of the average UK industrial electricity price.[23]

Because renewables are exempt from the tax, the immediate impact of the CCL was to make them effectively 0.43 pence per kWh cheaper. Not surprisingly, that increased demand for renewables. Green marketers, however, expected that the CCL would divert the limited green capacity to the nondomestic market and, in addition, increase the price they had to pay to obtain green capacity for resale. The *direction* of these effects, in the short term, was clear: The CCL made green generation more expensive, because of increased demand. The *magnitude* of this price increase, however, was unknown, and the long-term effects are unknown as well. The CCL, therefore, in the view of green marketers and providers, became a source of uncertainty and risk.

New Electricity Trading Arrangements

The New Electricity Trading Arrangements (NETA) is the UK's wholesale electricity trading system, which began operation in March 2001. Electricity trading is a complex and confusing world of its own. Fortunately, most of the gory details are not important for understanding its effects on the renewables market and thus are not discussed here.[24] A bit of history, however, is useful to understand the NETA.

A competitive electricity market needs some way for buyers and sellers of electricity to agree on prices and make deals. When electricity competition began in the UK in the early 1990s, a system called the Electricity Pool was established. Prices for electricity were set by a kind of bidding process. Generators offered to supply a certain amount of electricity at a certain price. For example, a power plant operator would offer to supply 100 MW for tomorrow 3:00 to 3:30 for 2.3 pence per kWh. The Pool would then arrange all the bids for a certain time period in order of increasing price, estimate the total amount of electricity needed for that time period, and then, starting with the lowest bid, select all those bids that were needed to meet demand. The price for electricity in that time period was set at the price of the highest successful bidder, and all successful bidders received that high price.

Although that system worked reasonably well, by the mid-1990s, it was increasingly criticized for being insufficiently flexible and for favoring large generators. It was then replaced by the NETA. The NETA is a wholesale market—a forum for buyers and sellers to set electricity prices. It allows bilateral contracts, which means that buyers and sellers can make direct, individual deals without having to make use of an official "pool" price. For example, a large electricity user (or a consortium of small ones) can make a direct deal with a generator to buy a certain amount of electricity for a certain time period. The electrons still flow through the national grid, of course, but the price is whatever the buyer and seller agree on.

What does this have to do with renewables and green markets? Under the old system, a renewable generator such as a wind turbine received the highest successful bid, even if their bid was lower. Under the new system, however, intermittent generators like wind turbines will have difficulty making bilateral deals, as they are dependent on intermittent resources such as the wind. This will likely mean that they will get a lower price for their electricity than they would have under the old system. In fact, wind generators saw a 25 percent drop in the prices they could get for their wind power in the first two months of NETA's operation relative to the same time period in the previous year.[25]

For green *marketers*, this was not necessarily bad. Lower wind wholesale prices mean either larger profits or reduced retail prices. But the *supply* of wind was the bottleneck, and lowering its value through NETA aggravated this problem.

Wind truly is an intermittent resource, and some argue that the old Electricity Pool system effectively subsidized wind by not adequately reflecting its intermittency in its price. Although NETA may result in what is closer to a "true" market price for wind, it did complicate life for green generators and for green suppliers. As with the CCL (see previous section), NETA increased uncertainty in the green market overall and further cowed already uncertain green marketers.

The Renewables Obligation

The renewables obligation (RO) is a requirement that electricity retailers (that is, companies selling electricity to end users) obtain an increasing percentage of their generation in the form of qualifying renewables. The RO began April 2002, with a renewable percentage requirement reaching 10 percent by 2010. (The RO is discussed in more detail in chapter 10).

The effects of the RO on the voluntary green market are similar to those of the CCL. In the long term, it will increase the supply of renewable generation, as it will increase the demand for renewables by forcing suppliers to purchase more renewable capacity than they would otherwise need. However, in the short term, it will divert renewable capacity from green markets to RO markets. The RO will also raise the price of renewable capacity in the short term, due to the increased demand. So the net effect, again in the short term, is to complicate life for green suppliers.

The RO may also require some slight shifts in marketing for voluntary green programs, but these effects will be small. The issue here is one of consumer perception: Will potential green electricity buyers see no reason to pay a surcharge for green electricity if suppliers are already *required* to supply some renewable electricity? There is a straightforward fix to this potential problem—ensure that the green product is a different, premium product, one that has either a higher percentage of renewables, or has "greener" renewables (for example, wind instead of large hydropower), relative to the RO electricity. In fact, several of the more sophisticated UK green marketers began to do so shortly after the RO came into effect.

What's Next for the UK Green Market?

One analyst noted in 2001, "The British green electricity market is in a stalemate of unclear government policies, timid suppliers, and under-enthused consumers."[26] The UK experience, although sobering, makes it clear that a successful green market requires smart marketing from the private sector and policy stability from government.

Notes for Chapter 4

1. Assuming a population of 60 million and an average household size of 2.3.

2. Ofgem (Office of Gas and Electricity Markets), "Competition Explained Main Page" (12 November 2002), from www.ofgem.co.uk (downloaded 25 February 2003).

3. As described on the London Electricity Web site. See www.london-electricity.co.uk/energy/business/green-tariff-1.html   (downloaded 25 February 2003).

4. As described on the Eastern Energy Web site. This program was later discontinued.

5. J. Lipp, "Policy Considerations for a Sprouting UK Green Electricity Market," *Renewable Energy 24* (2001), pp. 31–44.

6. This is called an RFP (request for proposal) in the U.S.

7. Clare Summers, Marketing Manager, Ecotricity, personal communication (29 March 2001).

8. Juliet Davenport, Director, unit[e], personal communication (20 February 2001).

9. As listed at www.foe.co.uk/campaigns/energy_and_climate/league_table.html in 2001.

10. UK National Audit Office, "Giving Domestic Customers a Choice of Electricity Supplier" (5 January 2001), London, The Stationery Office, p. 4.

11. MORI, "Experience of the Competitive Market," Research Study Conducted for Ofgem by MORI (January 2001), available at www.ofgem.gov.uk (downloaded 9 March 2003), p. 3.

12. MORI, "Experience of the Competitive Market," p. 3.

13. MORI, "Experience of the Competitive Market," p. 4.

14. MORI, "Experience of the Competitive Market," p. 3.

15. MORI, "Experience of the Competitive Market," p. 20.

16. MORI, "Experience of the Competitive Market," p. 24.

17. J. Lipp, "Appendix G: The UK Green Electricity Market—Is It Sprouting?," unpublished report, p. G9. Contact the author at judith.lipp@eci.ox.ac.uk.

18. Following quotes are from J. Lipp, "Appendix G: The UK Green Electricity Market—Is It Sprouting?" p. G10.

19. FT Energy, *Renewable Energy Report*, no. 25 (March 2001), p. 21.

20. Information from www.rspb.org.uk.

21. S. Boyle, "UK Green Power Grows but Still Lags Continental Markets," *Renewable Energy Report*, no. 42 (August 2002).

22. Or, as one wag noted, "It's too soon to take a 'wait and see' attitude."

23. Department of Trade and Industry (DTI), *Digest of United Kingdom Energy Statistics 2000* (London: The Stationery Office, 2000), p. 234.

24. Information on NETA can be found at www.ofgem.gov.uk/elarch/reta_contents.

25. "Ofgem Passes Buck on NETA as RO Enters Final Lap," *Renewable Energy Report*, no, 31 (September 2001).

26. S. Boyle and C. Henderson, "Green Energy: Withering on the Vine?" *Consumer Policy Review* (January/February 2001).

5

The U.S. Green Electricity Market: Lessons Learned

The first U.S. green power program began in 1993, and since then more than eighty electricity retailers have entered the U.S. green power business. Their successes and failures provide a wealth of lessons about how to make green power work—and about its promises and limitations as a tool to promote renewable energy. This chapter takes a close look at the U.S. green power market, with a focus on the detailed lessons that this experience provides. Results draw in part from interviews with green power program managers—the senior staff at energy companies with primary responsibility for making these programs work.

Findings

"It's really pretty basic marketing stuff," reported a utility marketing manager in northern California.[1] This comment nicely sums up the U.S. experience with green power. Energy retailers that applied essential marketing principles (such as branding and market segmentation) have generally done quite well. A fair number even sold out their green capacity. But others have failed, either because they were unwilling to invest in the necessary marketing and advertising or because they ran afoul of the shifting regulatory and industry restructuring winds.

The most successful U.S. green power programs have signed up about 6 percent of their residential customers, and some of these have sold out of green energy and are not accepting new customers. More typical, however, are sign-up levels of about 1 percent. Some green retailers in competitive areas are seeing rapid growth, with the leader in this market, Green Mountain Energy, signing up over half a million customers in seven states. New renewable capacity, mostly wind, is being built to meet this demand. As of February 2003, 980 megawatts

(MW) of new renewable capacity had been built expressly to meet the demand for green power, with an additional 430 MW planned.[2]

Clearly the green power market is composed of more than an obscure niche of dedicated environmentalists. Success in selling green power to date has been achieved largely through trial and error. Most of the marketing has been done by utilities, which are institutionally unsuited for marketing new products and services. As the understanding of green buyers advances, the marketing will become more sophisticated and successful, and penetration rates will continue to rise. Although the future of the U.S. green power market is very uncertain, one estimate forecasts that an additional 600 to 3,900 MW of new renewable capacity will be built by 2010 to meet green power demand.[3]

Regulatory Setting

The strange nature of electricity regulation in the U.S. sets the boundaries for the green power market, and so it's worthwhile understanding how that regulatory system works. (For a more detailed discussion of electricity regulation, see chapter 3). In the U.S., electricity markets are regulated and controlled at the state, not federal (national) level. As a result, each of the 50 U.S. states has set its own schedule and rules for restructuring: from California, which was the first to have a competitive electricity system in 1998 (with disastrous results, see box in chapter 3), to Georgia, which has shown little interest in opening up its electricity system. Most states fall somewhere in between.

The U.S. federal government plays a role similar to that of the European Union (EU) in Brussels: It sets overall goals and policy directions while focusing more on interstate than intrastate issues. The U.S. states, meanwhile, like the EU member countries, make final decisions about specific policies and schedules. Although the U.S. federal government has *considered* legislation that would require the U.S. states to liberalize their electricity systems, that legislation has not passed and furthermore is not likely to do so in the near future. As a result, the states are left on their own. In general, those states with higher electricity prices (which are generally those on or near the East Coast plus California) have been more likely to restructure, largely because of political pressure from large electricity users.

As of 2003, seventeen states had introduced competition or had set explicit schedules for doing so.[4] An additional six were planning to do so, but California's miserable experience convinced them to delay or cancel their restructuring plans. The remaining states had no explicit plans for restructuring.

U.S. green power markets are of two distinct types: those in restructured states and those in still-regulated states. These two markets are taking very different directions, and this chapter will, in most cases, treat them separately.

The U.S. Green Power Market Today

The first U.S. green power program was introduced in 1993 by the Sacramento (California) Municipal Utility District (SMUD), which offered photovoltaic power at a surcharge (that is, an additional charge on top of that charged for regular system power) of US$4 per month. This program sold out its initial capacity rather quickly, and other electricity retailers soon noticed its success and started programs of their own. The number of such programs grew dramatically: More than 100 regulated energy retailers were offering green power in 2003.

The green power market in competitive states has grown rapidly as well, as more and more states open up their markets. As of 2003, about 20 companies were offering green power products available in competitive states. Looking at both regulated and competitive markets, in 2003, about 40 percent of U.S. consumers had the option of buying green power.[5]

U.S. green energy products vary quite a bit in pricing, energy sources, and other attributes, but they can be grouped into four main types. For the regulated markets, most are contribution, wind, or mixed-fuel products:

- *Contribution programs* offer consumers the option of paying an additional amount on their energy bill that will be used to build new renewable generation. Typical of these are "round-up" programs, which ask consumers to round up their monthly payment to the next higher dollar amount (for example, from $37.84 to $38.00). These funds are then used to build new capacity, usually in the form of photovoltaic systems. The amount of money raised is usually modest, participation rates low, and resulting capacity built is minimal.

- *Wind programs* offer electricity from wind for a surcharge (that is, an additional charge on top of that for regular system electricity) of typically 2.5 US cents per kilowatt-hour (kWh). Most wind programs sell wind in blocks of 100 kWh per month, and consumers select the number of blocks they wish to purchase. Some utilities build their own wind capacity to provide the required amount of wind electricity; others contract for wind power from commercial wind farms. These programs have proven quite popular, and many utilities have sold out their initial commitment and have gone back to build or buy more wind capacity.

• *Mixed-fuel programs* offer renewables from a variety of sources, notably landfill gas and small hydropower. Most sell from existing generation, although some involve new capacity built specifically to supply green markets. Prices for renewable energy in these programs vary but are typically a surcharge of 1 to 4 US cents per kWh.

Green products offered in competitive energy markets are usually made up mostly of less-expensive existing generation (often small hydropower and landfill gas) supplemented by a small amount of "greener" wind and/or photovoltaics. Products in competitive states are usually priced relative to the default provider's rate (that is, the electricity rate charged by the original, formerly regulated retailer). Those choosing green power in competitive states typically pay an additional 1 to 3 US cents per kWh above the default provider's rate plus, in some states, a monthly flat charge of about US$5 per month.

Measuring Success I: Are Consumers Signing Up?

Is the U.S. green power market a success? The answer depends on how one defines success. This and the next two sections look at three distinct measures of success: participation rates, new renewable capacity, and profit.

The first question often asked of green power programs is, "What percent of consumers sign up?" This is an odd and somewhat inappropriate way to measure market size, because new product sales in most industries are evaluated on sales volume (units or gross revenue), not relative to an existing and largely incomparable market. For better or worse, however, program success is often defined by percent of market. Experience to date provides some data points that bracket the range of realistic sign-up percentages.

As shown in **Table 5-1**,[6] the most successful green power programs have signed up from 3 to 6 percent of their residential customers. Many of these programs have sold out of their green capacity and are building or contracting for more. Holy Cross Energy of Colorado, which sells wind power for $2.50 per 100 kWh per month, had no more wind power to sell when interviewed. Bob Gardner of Holy Cross noted, "We've got a waiting list...if we started marketing again, there are still some folks out there." Typical sign-ups are, lower than this 3 to 6 percent, averaging 1 to 2 percent. **Table 5-2**[7] shows the programs with the greatest number of participants, led by the Los Angeles Department of Water and Power (LADWP) with 73,000 residential sign-ups.

Table 5-1: U.S. green power programs with the highest participation rates, as of December 2002

Rank	Utility	Program	Participation rate (%)
1	Moorhead Public Service	Capture the Wind	5.8
2	Orcas Power & Light	Green Power	5.5
3	Los Angeles Department of Water and Power	Green Power for a Green L.A.	5.2
4	Holy Cross Energy	Wind Power Pioneers	4.9
5	Central Electric Cooperative	Green Power	3.7
6	Madison Gas and Electric	Wind Power Program	3.6
7	Sacramento Municipal Utility District	Greenergy, PV Pioneers	3.6
8	Preston Public Utilities	Wind Power	3.4
9	Cass County Electric Cooperative	Infinity Wind Energy	3.1
10	Cedar Falls Utilities	Wind Energy Electric Project	3.0

Table 5-2: U.S. green power programs with the greatest number of participants, as of December 2002

Rank	Utility	Program	Number of participants
1	Los Angeles Department of Water and Power	Green Power for a Green L.A.	73,000
2	Xcel Energy	WindSource	33,000
3	PacifiCorp	Blue Sky, Salmon-Friendly	20,000
4	Portland General Electric Company	Salmon Friendly and Clean Wind Power	20,000
5	Sacramento Municipal Utility District	Greenergy, PV Pioneers	19,000
6	We Energies	Energy for Tomorrow	11,000
7	Alliant Energy	Second Nature	7,000
8	Austin Energy	GreenChoice	7,000
9	TVA	Green Power Switch	6,000
10	Wisconsin Public Service	SolarWise, NatureWise	6,000

Are the 3 to 6 percent participation rates shown in Table 5-1 good or bad? For a new product that costs more than the product it competes against (system power, in this case) and that offers societal but not personal benefits, 3 to 6 percent is quite impressive. And this rate was accomplished as these retailers were just learning how to market green power. As consumer understanding of green power grows, demand will increase. Simultaneously, retailers will become better marketers and will also be able to shift marketing resources from education (telling consumers what green power is) to selling (convincing consumers to sign up). So it is likely that sign-up levels will increase further. (See also chapter 7 for a discussion of who buys green and why.)

Green power sign-up levels in competitive markets are similarly encouraging. In Pennsylvania, about one in five users who switched providers switched to a green product.[8] The California results are considerably more muddied. At first, virtually all of the 100,000 or so residential users in California who switched providers switched to green, but this was due in part to a generous government-sponsored credit, which made green power the same price as, or in some cases less expensive than, regular system power. Following the electricity market collapse in California, most of the green providers left the California market and returned their customers to the default provider.

Our interviewees generally set their goals and evaluated their progress in terms of either the number (or percent) of customer sign-ups or the amount of green electricity to which customers have subscribed. A somewhat surprising finding, however, was that most programs were not trying to maximize either sign-ups or green electricity sales but were, rather, trying to sell the green capacity they had in hand. Failing to sell the committed green capacity means that the costs of the unclaimed capacity would need to be covered from another source, while overselling meant disappointed customers on waiting lists. For example, "the specific goal was to have all the available energy spoken for by the time the project came on-line," according to Laura Williams of Madison Gas and Electric.

Our interviews also show, not surprisingly, that customer sign-up levels depend on how good the product is, how it is priced, and how well it is marketed. Williams commented, "We discovered that when our marketing campaign ended, public awareness quickly declined."

Dropouts: How Many?

It is clear that a green pricing program should ask for only minimal commitment from participants. One needs a "green pricing program that's very simple to get in and out of," suggested Bob Gardner of Holy Cross. This, however, raises the issue of dropouts. If people aren't committed up front, will they stick with the program? If not, how many will drop out, and when? After several years of experience with green programs, dropout rates are becoming clear: In the long term, they are at or just a bit above the background churn rate (that is, the turnover rate occurring due to moves, deaths, and changes in ownership).

Programs to date have seen a small flurry of dropouts in the first month or two after sign-up. This is due mostly to misunderstandings of how the program works. "People start seeing it on their bill, and they might drop out, because they didn't realize it worked like this," observed Laura Williams of Madison Gas and Electric. The extent to which this is a problem depends on how well the utility has explained how the program works and also what sort of programs you've had in the past. "Some that dropped out did so because they thought they were going to receive credit or a rate reduction...we took great pains to explain in our material that the cost of participation is over and above the usual electric bill. However, some customers missed that point," explained Russ Reno of Lincoln (Nebraska) Electric System. There may also be some disagreement within households about the program, suggested Williams, such as a husband saying, "My wife made the decision, not me, and I don't agree." Following this initial shakeout, programs have seen dropout rates at, or just a bit higher than, their normal churn rate. "The attrition we're seeing is mostly people moving outside our service area," noted Lori Clements-Grote of Fort Collins (Colorado) Utilities, a remark echoed by most of our interviewees.

Measuring Success II: Is New Renewable Capacity Being Built?

It's easy to lose sight of the primary societal goal of green programs: to increase the demand for renewable electricity and thereby create an incentive for construction of new renewables capacity. Have green programs actually driven new renewables construction? Analyses of new renewables has shown that the answer is yes. Green markets have led directly to the construction of over 980 MW of new renewables capacity, with another 430 MW planned (as of December 2002, see **Table 5-3**).[9] This capacity was built or planned expressly to serve green markets and does not include new renewables capacity built or planned to serve other markets (such as renewable portfolio standards). Note that the bulk of this new green capacity is in the form of wind.

Table 5-3: New renewable capacity built and planned to serve green markets, as of December 2002

Type	MW installed	MW planned
Wind	913	302
Solar	4.8	1.4
Small hydropower	8.6	2.0
Geothermal	10.5	49.9
Biomass	45.1	76.1
Total	982	431

Is this new capacity significant? As usual, it all depends on one's choice of denominator. Relative to the U.S. total non-hydropower renewable capacity of about 17 gigawatts,[10] the numbers in Table 3 are certainly encouraging.

Measuring Success III: Is Anyone Making Money?

The ultimate competitive market criterion for success is profit. Is green energy a profitable business? The answer so far is no, but for an unexpected reason: Most green energy retailers are not trying to make money. In fact, some are taking great pains to avoid making a profit on their green offerings.

Green retailers in noncompetitive markets are generally investor-owned or publicly owned utilities that offer green power as an alternative to their regular system power. Very few if any of these programs are priced or intended to make a profit—that is, to collect revenue that exceeds the program's costs. Interviews with program managers and utility management reveal a wider range of motivations for their green power programs, notably:

- *To gain experience in marketing a new product in advance of competition.* Many utilities expect to see competition in their states sooner or later, and they recognize that they will have to build marketing and new product–development skills. Green power is an opportunity to do so while still operating in the safe environment of a guaranteed and protected service territory.

- *To build a brand in advance of competition.* Utilities see green power as a way to build consumer loyalty and brand preference.

- *For public relations benefits.* Just about everyone likes renewable energy, although only a smaller number are willing to actually pay for it. Green power programs generally get very positive press coverage and community response, the value of which is hard to quantify but certainly not negligible.

- *To establish new political and marketing alliances.* The most successful U.S. green power programs are those that have comarketed with environmental advocacy groups, thereby improving the relationship between two groups that have traditionally been on opposite sides of many controversial issues. An unexpected but valuable benefit that green power has provided to utilities is building new and positive relationships with these traditional enemies—relationships that are especially valuable as state legislators consider restructuring legislation.

Notably missing from the list above is making money. In fact, some of our interviewees noted that they expressly tried to avoid making money from their green programs, as doing so would lead to all sorts of knotty problems with regulators and environmentalists.

The story from competitive states is different. Green power retailers in competitive states are mostly smaller start-up companies that are hoping to make money, although, so far, it isn't clear if they have managed to do so. The largest such retailer is Green Mountain Energy Company, which has 600,000 green customers in seven states.[11] Although this company does not release financial data, it is continuing to offer new products in more states and has attracted significant investment capital from BP and Nuon.

We Have Met the Enemy…and He Is Us[12]

Utilities don't think in marketing terms…it's a huge cultural struggle.

—Maggie Rogers, Sacramento Municipal Utility District

We interviewed a number of green power program managers to find out what they've learned about how to make green power a success. According to the green power program managers interviewed, designing and carrying out a successful green electricity program requires support and input from groups across the utility—from senior management to the call center to the billing department. This section presents some of the methods they've used to ensure this support.

It would be difficult to find an organization less suited to selling green power than the traditional regulated U.S. utility. Qualities associated with successful new product marketers—innovative, fast-moving, willing to take risks, able to make quick decisions—are at odds with the traditions of a regulated utility, which has long been rewarded for careful, slow, and risk-averse behavior. A number of U.S. green programs, however, have overcome these cultural predispositions and developed successful green power programs.

The Helpers, the Hinderers, and the Indifferent

Green programs are great for public relations. Russ Reno Lincoln (Nebraska) Electric System noted, "From a public relations perspective, it was one of the best things we've ever done." This benefit extends far beyond just those who sign up, because, according to Laura Williams, "most customers are in favor of renewables, whether or not they decide to buy." As a result, groups within the utility that are concerned with the utility's overall image—such as the public relations, corporate communications, or customer relations departments—are usually the biggest internal champions of a green program. The political and budgetary influence of these groups varies, of course.

In most cases, others at the utility are not actively opposed to the program but are far from convinced that it will work. Noted Williams, "Lots of [utility] folks throughout the company had a 'wait and see' attitude." The idea that customers will voluntarily pay more for electricity—and this often coming after years of focus within the utility on maintaining low rates—understandably produces a fair bit of cynicism. Although the initial argument for a green program needs to be convincing—as Gardner found, "We had to prove to management and the board that we could sell it"—few of our interviewees experienced active, vocal internal opposition.

Where apathy occasionally turns into real opposition is in the traditional generation group. A small number of our interviewees reported that the fossil-based generation groups felt threatened by the program, fearing that bringing in renewables would be seen as implicit criticism of fossil-based generation. Noted Chris Schoenherr of Wisconsin Electric Power Company, "There's a certain amount of 'don't say anything bad about my products.'" This is not a huge issue, but it does require some sensitivity in how the green resources are portrayed—"not that coal is bad, but that some customers just feel better about wind," reported Tim Seck of Great River Energy in Minnesota.

Overcoming the Somnolent Culture

The struggle against the lethargic traditional utility culture is real, but our interviewees found that evidence of customer demand for green power—from market research, from other utilities' success in selling it, and from high levels of customer sign-ups—did much to overcome the natural resistance to doing anything new. "There wasn't a whole lot of trust that this would be successful until we got real results," noted Williams. Recognizing this challenge up front and planning for it certainly helps, as she explained: "We knew the challenges would be in speeding up our decision making and keeping on schedule." Utilities often launch green programs for this very reason—that is, as a test case in offering a timely and market-driven product—and therefore it may be possible to enlist senior management support in driving decisions and progress.

Support from the Top

As with most efforts to do something new and innovative, green programs really do need support from the top. Reno explained, "When our new CEO came in, he recognized the growing demand among our customers for renewable energy. It might not have followed our ongoing commitment to use the lowest-cost energy, but our program was responsive to our customers. He saw the value in that and led the charge." It is often the case that senior management appreciates and values the public relations benefits of a green program and, in addition, understands the potential that green programs have for branding the utility. "Our chairman is the biggest fan and cheerleader," remarked Schoenherr, and this support is essential for getting the rest of the utility to support the program.

Call Center Coordination

For many utilities, green electricity is one of the first new products to be offered and, as such, it presents a new set of challenges for billing departments and call centers. This has led to tensions at some utilities, as these groups sometimes feel unprepared or understaffed. As Schoenherr observed, "every time we market the green program, it's going to increase call volume. If we do that at the wrong time, it can be tough for the call center." There is a relatively easy fix here—just be sure to coordinate your marketing efforts with the call center.

A few programs have run into problems with the call center not being trained to actually sell a product. For example, one interviewee noted that, when customers would call with questions about its green product, the call center would offer to send out literature rather than offering to sign the customer up for the product. Rudd Mayer of the Land and Water Fund advised, "Put time into training the people who answer calls about the program." A phased rollout or initial pilot program can help ensure that various internal systems, including the call center, are set up and trained as needed. This permits fine-tuning of these systems and helps avoid large-scale embarrassing problems.

Partnering with Environmental Groups: Worth the Effort

The partnership with environmental groups is key.

—Chris Schoenherr, Wisconsin Electric Power Company

The single largest indicator of success found in the review of U.S. utility green pricing programs was partnerships with environmentalists. Simply put, programs with active and positive partnerships of this kind are doing well, and those without are not doing well. It is clear that building and maintaining such a partnership is often time-consuming, difficult, and frustrating; but it is well worth the time and effort.[13]

Why Work with Environmental Groups?

There are five principal reasons to work with environmental groups:

- *Credibility.* The public support of an environmental group "gives consumers a sense that this is a legitimate green program," noted Gardner. Some utilities have a less-than-stellar environmental reputation, and an environmental partner can help overcome distrust of the program. This can also work at the regulatory level. Said Schoenherr, "If you can get the environmental groups on board and then you approach the commission, the potential opponents are already on board with you."

- *Access to potential buyers.* Members of environmental groups are often good candidates for participation in a green program. Madison Gas and Electric put sign-up forms in the local Sierra Club's newsletter.

- *Help with program design.* They are the target audience, and their reactions and suggestions can help design a successful program.

- *Marketing support.* Environmental groups "have taken brochures and had articles in their newsletter," said Lori Clements-Grote of Fort Collins (Colorado) Utilities. Reno pointed out that "they handed out information about our program at their booth at the State Fair." Environmental groups' marketing support has proven especially valuable in making commercial and industrial sales, because these users often buy green to improve their environmental image, and the presence of an environmental spokesperson at a sales meeting can help make the case.

- *Avoiding controversy.* Environmental groups can be powerful enemies, and working with them can avoid future conflicts. "It's better to have them with you than against you," suggested Schoenherr.

Understand Their World View

The relationship with an environmental partner will be stronger and more effective if an energy provider understands its goals and intentions. In general, environmental groups are primarily interested in environmental protection and see renewables as a way to reduce fossil fuel use. They are, of course, not motivated by developing the utility's brand or by establishing a green image for the incumbent utility. In addition, in many cases, environmentalists and utilities have historically been on the opposite sides of environmental issues. As a result, there may be little trust or faith in the utility's motives. Finally, there is some disagreement within the environmental community about green power. For example, some environmentalists are concerned that supporting the incumbent utility in establishing a green brand will make it more difficult for new green power suppliers to attract a critical mass of customers.

How to Work with Environmentalists

Our interviewees had several tips on how they've worked effectively with environmental groups. The first was to work with them from the beginning; getting the support and assistance of environmental groups requires contacting them early in the program design process. "The earlier you begin working with them, the better," advised Mayer. "Your product development may take longer because they're involved, but you're going to end up with a better program, a better product, and a lot more support." This may sound like a burden, but in fact it can be a great help. "The environmental groups actually helped us design the product, the pricing, and the marketing," reported Laura Williams. These groups know green energy's target audience better than anyone.

The second piece of advice from our interviewees is that environmental groups have a very strong preference for new renewables, rather than existing supplies—in fact, their support is probably contingent on this. "United Power started a program selling wind from existing resources, and they got criticized like crazy by environmental groups," noted Seck.

The third and final finding was that early and frequent communication about goals and mission is essential. Andy Sulkko of Public Service Company of Colorado noted, "Everybody needs to realize where everybody's coming from right from the beginning...having a narrow and agreed-upon mission and objective."

The U.S. green power market is new and still suffers from inept marketing, consumer ignorance, and lack of faith from both utilities and some environmentalists. Despite these barriers, it has managed to garner a nontrivial market share, which will certainly grow as the marketing sharpens and consumer understanding improves. The shrewd use of green power programs, along with other policy options, will do much to help renewable energy contribute to world electricity supply.

Notes for Chapter 5

1. P. Komor, "Making Green Electricity Programs Work: The Experts Speak Out," E Source Green Energy Series GE-5 (September 2000). This quote, and subsequent information from interviews, is from this report.

2. L. Bird and B. Swezey, National Renewable Energy Laboratory (NREL), "Estimates of Renewable Energy Developed to Serve Green Power Markets in the United States" (February 2003), from www.eere.energy. gov/greenpower/new_gp_cap.shtmlat (downloaded 22 April 2003).

3. R. Wiser et al., "Forecasting the Growth of Green Power Markets in the United States," NREL/TP-620-30101 (October 2001), p. vi.

4. Data from U.S. Department of Energy, "Status of State Electric Industry Restructuring Activity as of February 2003," from www.eia.doe.gov/cneaf/ electricity/chg_str/regmap.html (downloaded 23 April 2003).

5. B. Swezey, "Introduction to Regional Green Power Marketing Reports," presentation at the Seventh National Green Power Marketing Conference, Washington, D.C. (30 September 2003).

6. U.S. Department of Energy, "Top Ten Utility Green Pricing Programs as of December 2002," available at www.eren.doe.gov/greenpower/topten.shtml (downloaded 4 April 2003).

7. U.S. Department of Energy, "Top Ten Utility Green Pricing Programs as of December 2002."

8. B. Swezey, "An Overview of Green Power Marketing in the U.S.," presentation at First European Conference on Green Power Marketing, St. Moritz, Switzerland (28–29 June 2001).

9. L. Bird and B. Swezey, "Estimates of Renewable Energy Developed to Serve Green Power Markets in the United States."

10. U.S. Department of Energy, from www.eia.doe.gov/cneaf/solar.renewables/ page/rea_data/tablec14.html.

11. See www.greenmountain.com.

12. Pogo Possum, as quoted by Walt Kelly (April 1971). See www. nauticom.net/www/chuckm/whmte.htm.

13. For an excellent case study of a successful utility–environmental group cooperative effort to sell wind power, see R. Mayer, E. Blank, and B. Swezey, "The Grassroots Are Greener," Renewable Energy Policy Project Research Report No. 8 (June 1999), Renewable Energy Policy Project, Washington, D.C., from www.repp.org.

6

The Dutch Green Electricity Market: Lessons Learned

The Netherlands' green electricity market is the most successful in the world, with participation rates far above those seen in other countries. This is a remarkable fact given that the Netherlands is a heavily urbanized country with limited renewable energy resources and a ready availability of inexpensive natural gas from its Groningen natural gas fields, which are among the largest natural gas fields in Western Europe.

In the Netherlands, as in the UK case study, green markets have proven to be unpredictable. The factors determining success or failure can be subtle and unexpected. This chapter tells the story of the Netherlands green market, focusing on what worked and why.

History and Current Status

The Netherlands' green electricity market began in 1995, with a pilot program run by the Dutch utility PNEM. By the following year, a modest total of 35 giga-watt-hours (GWh) of green electricity had been sold. Green sales grew briskly from there, but by 1999 were still at well under 1 percent of electricity sales. In September 1999, an imaginative marketing campaign by the environmental group WWF (World Wind Fund for Nature) led to a dramatic leap in green sign-ups. By 2000, about 3.5 percent of Dutch households were specifying green electricity for their home use.[1]

At this point, the Netherlands' green market became intertwined with the restructuring of the electricity market. As of 1 July 2001, residential consumers were given electricity choice—but for green products only. In other words, residential electricity users could stay with their default provider, or they could choose another provider's green electricity product. They were not able to switch to another provider's brown product. As discussed below, this led to a flurry of innovative marketing of green electricity and a rapid increase in sign-ups.

By May 2003, green sign-ups reached 1.8 million—or 26 percent of Dutch households.[2] To put this in perspective, the more successful U.S. green programs have seen penetration rates of 3 to 6 percent—but this 26 percent Netherlands number is for the *entire country*, not just for a selected program. In fact, some Netherlands green suppliers stopped marketing because they couldn't get enough green electricity to sell.[3]

Green electricity sales in the Netherlands were not restricted to the residential sector: In 2001, about one-fourth of Dutch green electricity sales were to nonresidential buyers.[4] Large Dutch nonresidential green buyers include the national Dutch railway, the Amsterdam municipal water company, and the Utrecht city government. In addition, the Dutch government plans to purchase green power to meet a portion of its electricity needs as part of an effort to reduce carbon emissions.

Why Is This Market So Successful?

So why is the Netherlands' green market so successful (as measured by market penetration)? There are four principal explanations for its success, listed here in order of importance:

- Due to heavy taxes on fossil-based electricity, green electricity is about the same price as non-green electricity.

- The green market opened first, creating a competitive green market in advance of overall competition.

- Dutch companies have used creative marketing techniques to promote green electricity.

- In addition to fossil fuel taxes, the Dutch government has several other aggressive policies that support renewable electricity.

Price

Most green electricity products in the Netherlands cost about the same as regular, brown electricity. This is due to the Ecotax, also called the regulating energy tax (REB in Dutch). This is a per-kilowatt-hour (kWh) tax, which is paid directly by electricity users in proportion to their consumption. The amount (level) of this tax is considerable, especially in contrast to other similarly structured taxes, such as the UK's climate change levy (CCL) of 0.6 US cents per kWh. In 2001, the Netherlands' Ecotax was 4.8 US cents per kWh (**Table 6-1**).[5]

Table 6-1: The Netherlands' Ecotax

Year	Tax (US cents/kWh)
1996	1.2
1997	1.2
1998	1.2
1999	2.0
2000	3.3
2001	4.8

Note: Conversion of 1 NL guilder = US$0.4 assumed.

Qualifying renewable electricity, however, has been exempt from the Ecotax since 1998. In other words, electricity users pay an additional 4.8 US cents per kWh (2001) on nongreen electricity only. The end result is that green electricity has become effectively less expensive as the Ecotax has increased, and by 2001 most green electricity products offered in the Netherlands were typically about the same price as regular, brown electricity.

One obvious reason for the success of the Netherlands' green market is that it is free. One would think that if green is about the same price as brown, then of course people would buy it, because they get the green attributes at no cost.

A look at other markets, however, shows that this isn't quite accurate. There have been several UK green electricity products, notably those offered by Ecotricity and RSPB/Scottish and Southern, that were "free" (meaning that users would pay no more for them than for regular, brown electricity) but saw only moderate take-up. Similarly, when the California residential electricity market opened up, many green products were offered at or in some cases below the price of brown, and many saw very low sign-up levels. In contrast, some of the more successful U.S. programs have been those charging premiums of 2.5 US cents per kWh. So price is important, certainly, but it's not the entire explanation for the success of the Dutch green market.

Green Opened First

As noted above, the Dutch restructuring plan involved a very innovative step: Green opened first. Although residential electricity consumers did not have full choice until 2004, as of 2001, they were able to switch to a competing provider's green product. This led to a significant increase in green sign-ups for two reasons.

First, retailers saw this early, limited opening of the Dutch domestic market as their opportunity to build brand awareness and increase their customer base. So they started marketing their green products even before 1 July 2001 and continued to pay considerable attention to the green market after that.

Second, there is always a small but highly motivated group of electricity users who hate their default supplier. They may have had a billing or other dispute with the supplier; they may have been incensed by an article in the press; or they may have heard a story from a friend about the supplier. Whatever their motivation, they will switch from the default supplier as soon as they can. As the only option from 1 July 2001 was switching to green, the green market got the angry and disaffected market niche.

Smart Marketing

The marketing of green electricity in the Netherlands is notable for several reasons: its success in attracting sign-ups, its imaginative approach, and its involvement of various stakeholders. For example, in September 1999, the environmental group WWF launched a major marketing campaign to get people to sign up for green electricity. Using the phrase "Don't let the North Pole melt...choose green tariffs," the WWF attempted to link protection of the northern Arctic ecosystem to green electricity. Global warming is thought to contribute to sea-level rise and Arctic ice melting, which could, among other effects, reduce polar bear habitat. Greater use of green electricity could delay or reduce global warming.

The "Don't let the North Pole melt" effort was launched by 2,000 volunteers, who laid a 270-km-long (170-mile) green ribbon along the Dutch coastline. This ribbon was intended to represent concern over rising sea levels. This was followed by a two-week publicity campaign. Awareness of the green tariffs increased significantly, and in the following four months, an additional 44,000 new green customers signed up—a 38 percent jump in participation.[6]

The period after July 2001, when green choice was introduced, saw an impressive spread of innovative marketing of green products:

- Echte Energie, a Dutch online supplier of green energy, sold green energy through health food stores. Those signing up in the stores received a US$18 gift voucher.

- Another new green provider, Caplare, targeted the ethnic market, offering bills and marketing information in Turkish, Arabic, Vietnamese, and Chinese. It also provided a voucher for US$10 of international phone calls with each sign-up.

- Shell began marketing green energy through its 700 gasoline retail outlets in the Netherlands. It included a price guarantee, whereby customers committing to a year of green electricity were guaranteed a fixed price for the year.

- The Greencab company in Utrecht, the Netherlands, began offering taxi service using electric vehicles powered by green electricity.

The Policy Panoply

The Dutch Ecotax of 4.8 US cents per kWh (2001) is the policy giving the greatest boost to the Netherlands' green market, as it makes green the same price as brown. But there are a number of additional policies in place that help the green market and are responsible in part for its success. Many of these are on the supply side—that is, intended to help generators of renewable electricity, but they ultimately affect users as well, by reducing prices and increasing availability. Significant policies include:

- Tax exemptions for those investing in green funds,

- Accelerated depreciation for certain technologies,

- Tax credits for certain technology investments,

- Direct support payments to renewable generators, and

- A green certificate trading system (see chapter 11).

Constraint: Can't Build New Capacity

The Netherlands, like the UK, is challenged in building new green capacity due to local planning opposition. This is most relevant to wind projects, as the best wind resource is usually found near the shore and on tops of hills—both areas where open views are prized. This is, not surprisingly, a controversial and politically charged issue. The Green party spokesperson noted, "Wind turbines are more important than a nice view,"[7] and some are pushing for legislation that would require municipalities to install wind turbines. There have also been conflicts among landowners—those with turbines on their land often do receive some sort of rent; however, those on neighboring land suffer from the visual intrusion but don't see any of the associated revenue.

Several efforts have been made to ease and shorten the planning process. For example, a campaign called Space for Wind Energy provides information on how wind energy collection and agriculture can both take place on the same land. Studies are underway or completed on putting turbines near railway lines or dikes. The Ministry of Economic Affairs is supporting efforts to find acceptable sites for wind turbines in the Friesland and Noord-Holland regions.[8] There has also been a renewed interest in biomass, due in part to a belief that it is easier to site biomass plants.

A more aggressive—and controversial—solution to renewable supply shortages is to import renewable electricity from other countries. New rules issued by the Dutch Minister of Economic Affairs as of January 2002 allowed imported renewable electricity to be sold as green if it met certain criteria. The rules for imported hydropower were much stricter than for renewables produced from other fuels.

Administrative Challenges

The Dutch green market, although clearly successful (if one defines success as the percentage of users signing up for it), has had a number of administrative challenges along the way. It was originally set to allow green choice in April 2001 but was delayed until July 2001 because of problems in setting up green certificate and tax exemption systems. There is ongoing controversy over the role imported renewable electricity should play in the Dutch green market. Domestic renewable capacity is insufficient to meet the needs of all green customers, but many question the appropriateness of offering generous subsidies to imports, which may have already benefited from subsidies in the originating country. Also, in the Netherlands, as in all European Union (EU) countries, there is always concern about green policies running afoul of EU rules—especially those related to "state aid." (EU state aid rules prohibit countries from subsidizing their domestic industries.) A 1998 European Commission (EC) ruling found the Dutch Ecotax system did not violate state aid rules, but the Ecotax has been increased since then, and EC approval is awaited.

What's Next for the Dutch Green Market?

The Dutch economic philosophy has long tilted toward market-based solutions to public policy. The Dutch green system—using taxes and consumer choice, rather than regulation, to promote renewables—is a perfect example. Its success, however, is and will continue to be limited by supply constraints. Key uncertainties right now are the role of imported renewable electricity and the related issue of green certificates. A widely accepted and recognized EU-wide green certificate trading system appears to be several years away at best (see chapter 11), yet without it there will continue to be controversy over imported renewable electricity.

The green market in the Netherlands demonstrates that it is possible for green power purchases to achieve a significant market share—the Dutch market is already over 25 percent and is continuing to grow. This was, however, not a "pure" market outcome—the tax exemption and the early opening of the green market to choice had much to do with its success. And supply constraints are a continuing problem, demonstrating the need to keep policy focus on both the demand and supply sides of the renewables market.

Notes for Chapter 6

1. K. Kwant and W. Ruijgrok, "Deployment of Renewable Energy in a Liberal-ized Energy Market," Novem (undated), from www.novem.org (nlre.pdf), p. 12.

2. From www.greenprices.com/nl (downloaded 10 June 2003).

3. K. Kwant and W. Ruijgrok, "Deployment of Renewable Energy," p. 8.

4. From www.greenprices.com/nl (downloaded 13 April 2001).

5. Netherlands Energy Research Foundation, "Energy Market Trends in the Netherlands 2000," available at www.ecn.nl (downloaded 23 January 2002), p. 26.

6. From www.greenprices.com/nl (downloaded 13 April 2001).

7. As quoted in Nuon, "Environmental Report," from www.nuon.nl/about (downloaded 14 March 2001), p. 13.

8. Netherlands Ministry of Economic Affairs, "Renewable Energy in Progress 1999," from www.minez.nl (downloaded 11 April 2002), p. 45.

7

Understanding Green Buyers: Why They Choose to Pay More for Electricity

The green market has been in existence long enough to yield useful data on buyer descriptions and motivations, and it paints a surprisingly nuanced and revealing picture of who's buying and why. Residential green buyers' motivations are quite different from those of commercial green buyers, and so this chapter discusses each separately.

Who Are the Residential Green Buyers?

As of 2003, residential green power purchasers numbered some 1.4 million in Europe[1] and 800,000 in North America.[2] The common—and misleading—belief is that these buyers are "dark greens"—individuals with very strong environmental beliefs whose environmental commitment leads them to willingly pay a premium for renewable energy. The truth, however, is both subtler and more encouraging for green markets. Green buyers are not all dark greens, but rather are surprisingly diverse and eclectic. It is revealing to characterize green electricity buyers several ways:

- Based on *demographic* indicators, such as income, education level, or gender.

- Based on *attitudinal* or *value* indicators, such as environmental activism or political affiliation.

- As *economic* actors, looking for a return on their green investment.

None of these perspectives perfectly characterizes green buyers, but all three together paint a nicely detailed picture of just who's buying green energy and why.

Demographics of the Green Residential Buyer

A large survey of green energy buyers and nonbuyers across North America was undertaken in 2001 and 2002.[3] This survey took data from 2,800 residential electricity users and is the most comprehensive and useful to date. The survey asked participants a number of questions related to political beliefs, views on environmental issues, and other possible indicators of green buying behavior, as well as standard demographic indicators. The results were then broken down into green buyers and nonbuyers. Results from this survey confirm some commonly held ideas about green buyers but also show that a few beliefs are incorrect.

Indicators showing the greatest difference between buyers and nonbuyers were political persuasion and support of environmental groups. Almost two-thirds of green buyers identified themselves as "liberal" (as opposed to just 38 percent of nonbuyers), and just under two-thirds of green buyers were supporting an environmental group (as opposed to just 28 percent of nonbuyers). Education level was also a strong predictor, with green buyers much more likely to hold advanced degrees. Green buyers are also more likely to be politically active.

As the study's authors noted, environmental activism was a surprisingly strong indicator. "(We were) not surprised to find that participants were more likely to have knowledge of, and interest in, the environment. We were, however, surprised by how powerful an indicator environmentalism proved to be. And when we asked participants to explain why they joined the green energy program, they most often gave explanations couched in environmentalism: 'It's an opportunity to do something for the environment'; 'Because I think we should protect the environment and it's convenient'; and simply, 'For environmental reasons.'"[4] Also notable is what *didn't* show up as significant. Neither income nor gender showed much variation between buyers and nonbuyers, in contrast to the usual belief among marketers that higher income groups and women are more likely to buy green.

Market Segmentation

A different approach to understanding just who buys green power is the use of market segmentation. The marketing profession uses this approach to separate consumers into similar classes, or segments, which can then be targeted by segment-specific advertising. This is essentially a variation of the diffusion of innovations theory, first presented by E. Rogers in 1962, which portrays new technologies and practices as following an S-shaped adoption curve. The first to adopt it are "innovators," followed by "early adopters," and so on (**Figure 7-1**).[5]

Figure 7-1: Rogers' innovation diffusion curve

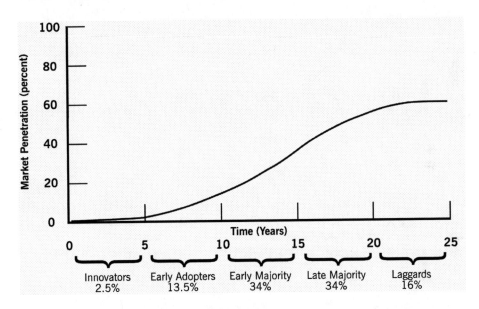

Market segmentation is fast becoming the preferred method for marketing green power. One green power program manager noted, "The more you can target a group of people, the better your response is going to be."[6] Similarly, another found that "one of the real keys to success is understanding who would be interested in the product."[7] Targeting marketing messages both increases yield and reduces costs, as resources aren't wasted sending messages to people who are probably not interested. Of course, segmentation comes at a cost, so the generally higher yield rate of targeted marketing must be traded off with its increased cost.

An analysis of the Western European residential green power market[8] defines the first round of green buyers as both "dark greens" and "innovators" (see Figure 7-1). According to this analysis, beyond this niche are the "light greens," who differ from the first group in several ways:

- They put more importance on individual, rather than just social, benefits of their purchases.

- They rank other features (such as price) as important as greenness when making purchase decisions.

- They have less knowledge about energy and the environment and thus need more basic information.

Of the many consumer segmentation schemes available, PRIZM has emerged as particularly useful for identifying green energy buyers. The 2001/2002 North American survey identified five PRIZM clusters as containing those most likely to participate in a green pricing program (see **Tables 7-1**[9] **and 7-2**[10]).

Table 7-1: PRIZM green indices

Cluster name	Green index[a]
Urban Achievers	4.5
Blue Blood Estates	3.5
Winner's Circle	2.5
New Empty Nests	2.4
Pools and Patios	2.2

Note: (a) Green index is the normalized ratio of participants to nonparticipants in that segment, and is thus a measure of the relative likelihood of a member of that segment buying green.

Table 7-2: PRIZM definitions

Cluster name	Definition
Urban Achievers—Mid-level white-collar urban couples	Often found near urban public universities, these neighborhoods are ethnically diverse with a blend of youth and age. Single students mix easily with older professionals who work in business, finance, and public service. Affluence is "middle" among the clusters. Age groups: 25–44, 65 and older.
Blue Blood Estates—Elite, super-rich families	Established executives, professionals, and "old money" heirs live in America's wealthiest suburbs. They are accustomed to privilege and live luxuriously. One-tenth of this cluster are multi-millionaires. Age 45–64.
Winner's Circle—Executive suburban families	These "new money" families live in expensive mini-mansions in major metropolitan suburbs. They are well-educated executives and professionals who are married with teenagers. Big producers and big spenders. Winner's Circle families enjoy globetrotting. Age 45–64
New Empty Nests—Upscale suburban fringe couples	Hard work in professions and industries has rewarded New Empty Nesters with the affluence that comes from double incomes. Most of these married couples are in their "postchild" years, are far more conservative than Young Influentials (a cluster of high-tech managers and professionals), and live in the Northeastern and Northwestern U.S. Affluence is upper middle, ranked 15 among the clusters. Age 45 to post-65.
Pools and Patios—Established empty nesters	Empty-nester executive and professional couples are living the good life in their "post-child years." Their dual incomes support rich active lives filled with travel, leisure activities, and entertainment. Many live in the densely populated Northeast corridor of the U.S. Affluent, in top 10 among clusters. Age 45 to post-65.

Although these market segmentation schemes can seem rather imprecise and ad hoc, there is more value here than one might think. Most U.S. households are *already* classified into one of the sixty-two PRIZM clusters based on zip codes. Most marketing, from direct mail to television advertising, already makes use of PRIZM-like market segmentation techniques. It's only the energy industry, with its regulated-monopoly legacy and attendant indifference to consumer preferences, which has been slow to recognize the value of such methods. Greater use of these techniques will clearly lower marketing costs and increase green power market penetration in the future.

Interviews with program managers reveal further subtleties about who buys green and why. An interesting nuance is the imperfect overlap with energy-efficiency interest (which many utilities have information on from past efficiency programs). One green program manager noted that green buyers are "generally, people who care about energy conservation, but not to the extent that they think about it only as a way to save money."[11] Those whose interest in energy conservation is driven by financial, rather than environmental, considerations are not good prospects for green electricity. Another attitudinal predictor often raised was community activism. "Customers involved in our green program are very involved with the community," observed another manager.[12]

Green Buyers as Economic Actors

A different perspective on green buyers is economic—that is, understanding buyers in terms of what they are willing to pay and what they expect to receive for their investment. Most utility green electricity programs are using a block pricing structure in which buyers pay a surcharge (typically US$2.50 per 100 kilowatt-hours [kWh] per month) for green electricity. This allows a low buy-in point and, as a result, "the price point is low enough that it's not really an economic decision," said a green program manager. This isn't to say that price doesn't matter—"price is more relevant to some consumers than to others," he noted—but, in general, a low buy-in point means that the buying decision is determined mostly by concerns other than price.[13]

There is, however, a clear need to make the consumer's return on investment as clear, tangible, and direct as possible. That can be achieved through environmental protection or the construction of new renewables. Explained one green power program manager, "You want to keep in touch with your customers, because the way you sell this product is through the environmental benefit that they are buying. If you're not telling them what has been accomplished by their participation, they're going to lose interest." Said another, "One key thing we learned is we needed to make the product as tangible as we could. Our customers wanted to know things like, 'How many new wind turbines have been built because of our dollars?' We've created a newsletter that goes out to all customers a few times a year that will talk about things like how many people are on the program and how much emissions reduction their contribution is providing."[14]

Although these various demographic and attitudinal descriptors do capture many green energy buyers, several of our interviewees noted that these measures are far from perfect. "It's not just environmentalists or suburban couples…we're finding it's lots of different people," observed a program manager. One surprise was the popularity of green electricity programs in rural agricultural areas. Noted another, "We know that farming is a tough business, but the reality is that we're getting a good response from those people."[15] On the other side, several programs have struggled to sign up college students. Although their values are largely aligned with green programs, they are transient, often don't have utility bill–paying responsibility, and usually have very low incomes.

What Do Residential Green Buyers Want from Their Green Purchase?

Green power is an amorphous product. Almost identical to regular system power, its only obvious distinguishing characteristic is that it usually costs more. However, considering that consumers are willing to pay that premium, in their eyes it clearly does differ in other ways as well. So just what is it that they think they are buying? For residential consumers, the purchase can be seen as an expression of environmental values. As more and more green products come on the market, the details of how consumers define these values, and translate them into product preferences, are becoming clear.

Green products themselves can vary on several attributes, notably:

- generation source (wind, solar, etc.);
- generation location (local, regional, national, etc.);
- generation vintage (planned, newly built, existing); and
- price.

Although consumer preferences change continually, experience to date suggests that perceived "greenness" shows the patterns summarized in **Figure 7-2**.[16] The "purest" forms of generation are seen to be wind and solar. Hydropower is seen as somewhat less green, followed by waste-fired generation. These preferences are usually based on limited information and thus are easily changed—for example, a newspaper article about birds killed by wind turbines can easily shift the ranking. Landfill gas preferences are not yet clear, as few consumers know what that is. Similarly, there is some evidence that consumers prefer newly built renewable generation over existing. This is clearly the preference of the environmental community, for obvious reasons, and their views are often shared by consumers as well.

Figure 7-2: Consumer perceptions of "greenness"
Consumers generally associate greater "greenness" with generation that is new, local, and wind or solar.

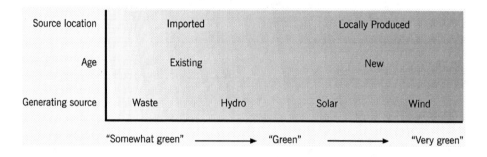

Pricing is an interesting attribute. The general lesson so far is that consumers care less about price than do most in the energy industry. This is again a legacy of the electricity industry's regulated past, where price was the only variable and was the key instrument of regulatory control. As noted above, price does matter; however, buying green energy is not a financial decision but a value-driven one.

That said, two pricing-related points were consistently noted by U.S. green energy program managers: simplicity and a US$5 per month critical price point. Residential consumers in particular don't want to spend time and effort to understand a complex pricing program. Furthermore, they're suspicious of any pricing structure that they don't see as transparent and obvious. The simpler the price, the better. (This is well illustrated by trends in pricing for other goods and services, such as long-distance telephone service and Internet access. Prices started out complex and cost-based yet moved quickly toward simple and consumer-friendly.) Five dollars per month was seen as a green energy price point that was acceptable to consumers.

Nonresidential Green Buyers: Who and Why

One early surprise for green power programs was the demand for green power from nonresidential (nondomestic) electricity users, such as municipalities, churches, and retail stores. Even in the absence of any serious marketing to these users, a significant number have specified green power for their electricity needs. In 1999, for example, more than one-fifth of the green power in California was bought by commercial and industrial users.[17] Some sales have been huge: Swedish Railway has purchased 1.2 billion kWh per year of green energy, and the Authorities Buying Consortium (a purchasing agency for local governments in Scotland) has purchased 112 million kWh per year of green power from Scottish-Power.

Why would these users, especially those seeking to make a profit, willingly pay more for electricity? What are they looking for in return for their green purchase? Are they just the commercial-side equivalent of "deep greens," or is this a significant and enduring market? Survey data and case studies show that, for nonresidential green electricity buyers:

- Price matters—but less than one might think. For small deals, such as with small retail stores, a modest price premium is acceptable as the buying decision is, as in the residential sector, value-driven. For larger deals, innovative financing and bundling (with energy efficiency, for example) are needed in order to keep costs down.

- Even for large companies and government/public agencies, decisions to buy green are driven by moral/ethical beliefs, not by business or economics. Although economic justifications are handy, they are not the main force behind these decisions.

- Large deals require an internal champion—someone within the organization who is committed to seeing the deal go through.

Motivations for buying green vary considerably by the type of buyer, so this discussion distinguishes the three major types of nonresidential buyers: for-profit firms, public (government) agencies, and nonprofit organizations.

For-Profits

For-profit firms buying green power range from small retail stores buying green to express their environmental values (a purchase decision similar to that of residential buyers) to national companies looking to add substance to their environmental commitment statements. The diversity of these firms is illustrated by **Table 7-3**,[18] which provides just a few examples of for-profits buying green power.

Table 7-3: Examples of for-profit organizations buying green power

Example 1	Kinko's, a provider of document production, copying, and business services, signed an agreement in November 1999 with a renewable energy provider to supply the load of all 75 branches in California with 100 percent renewable power.
Example 2	On Earth Day 1998, Toyota Motor USA announced that it would purchase enough green energy to cover the entire load of the company's sales and marketing arm. Toyota purchased 12 megawatts of 100 percent green energy, at a US$1 million price premium per year. (See "Case Study: Toyota" in text.)
Example 3	The Los Angeles Dodgers baseball team agreed to buy 7.3 megawatt-hours per month of green energy to be used to cover part of the Dodger Stadium load.
Example 4	Proctor and Gamble UK purchased 121 million kilowatt-hours per year of green energy—half their total consumption—from London Electricity plc.
Example 5	The accounting firm KPMG UK purchased 32 million kilowatt-hours per year of green electricity from London Electricity plc.

Surveys and interviews with for-profit companies buying green power show that the principal—but certainly not the only—motivating factors for buying green are internal organization values and civic responsibility (**Figure 7-3**).[19] Although these data are based on self-reports and thus may be subject to some bias, they suggest that altruistic concerns do drive behavior. This is supported by survey results showing that most companies have done very little to publicize their green power purchases.[20]

Ranked just behind organization values and civic responsibility was employee morale (defined in the survey as "employees feel more pride in an organization that is giving back to the environment"). This surprisingly high ranking suggests that sign-ups are not just based on management preferences, but that employees' preferences play a role as well.

Figure 7-3: Altruistic factors drive green power purchases
Shown here are the results of a survey of businesses buying green. Interviewees were asked about each factor's influence on their decision to buy green power. A mean response of 5 = "very important" and 1 = "not important."

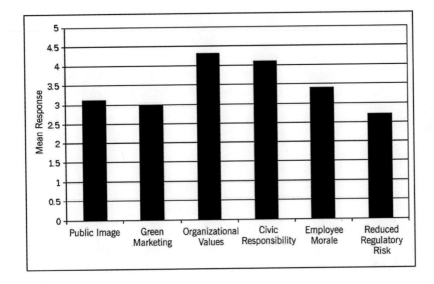

The motivation factors shown in Figure 7-3 are aggregated across a variety of for-profit companies. A breakdown of these factors by company size, however, revealed an interesting variation. Large companies tend to rank public image more highly, suggesting that altruistic factors alone are not enough to sway a large company into buying green. Smaller firms, in contrast, tend to behave more in line with their founder's or principal's personal beliefs.[21]

It's useful to recognize that these for-profit companies are, in many cases, the early adopters—willing to seek out and pay for a premium and relatively unknown product. It's therefore no surprise that internal/altruistic motivations rank high. As the green power market develops, it's likely that this will change. As green power becomes a more recognized product, for-profit companies may increasingly see a competitive advantage in green power.

Surveys and interviews also reveal how for-profit companies make the actual decision to buy green, and what they look for in a green power product. A consistent factor is the importance of an internal champion—someone in the organization who sells the green product internally. According to a business development manager at Green Mountain, the largest green power marketer in the U.S., "there is almost always a single champion—someone who takes the lead and really makes it happen."[22] For small organizations, this champion is in most cases the owner, director, or CEO; for large companies, the facilities staff is in some cases the champion.[23]

Surveys have found that the most important criteria for-profit companies look at in a green electricity product are, in order, the percentage of renewable energy in the product, the degree to which it includes new renewables (as opposed to existing capacity), and the type (generation source). Coming in fourth, surprisingly, is price.[24] As in the residential sector, it appears that price is less of a concern than is commonly thought.

Case Study: Toyota[25]

On Earth Day 1998, Toyota Motor USA announced that it would purchase enough green energy to cover the entire electricity needs of the company's sales and marketing groups, at a US$1 million price premium per year. At the time it was announced, this was the single largest purchase of green energy by any U.S. company.

Jim Cooke, who at that time was the national manager of real estate and energy affairs for Toyota, noted that a lot of things had to fall into place before making the green purchase, and the overall process was time-consuming. The key driver was corporate policy.

Purchasing green energy fits well with Toyota's corporate policy. "One of Toyota's 'Guiding Principles' is to exist in harmony with the earth," says Cooke. "We take that principle very seriously, and taking the proactive measure of purchasing green energy was one way for us to demonstrate our commitment. We believe that the use of green power is a statement and challenge to our business partners, our competitors, and the rest of corporate America that it is important for all of us to take responsibility for our environment."

But even with a strong environmental focus, the decision to purchase green energy was anything but easy for Toyota. "You must remember that our core business is selling motor vehicles. That is where our senior management focuses its attention, and I think rightly so." What that meant for Cooke was that he and his colleagues spent considerable time providing basic education about the energy market and the green energy market specifically. Questions were often focused on the overall reliability of green energy and whether the lights would stay on when renewable resources like wind and solar weren't available. "This may seem like simple stuff for you, but it's important to remember that these people have other things on their mind and don't understand energy markets."

Cooke says that the experience of trying to sell green energy to others at Toyota has convinced him that large commercial customers should be viewed as "multiheaded customers. People like me have their own customers within the company. There are many people weighing in, and anything could go wrong at the last minute." Shepherding the purchase of green energy through senior management proved to be difficult right up to the very end. In fact, the very day before Toyota was to announce the purchase, the CFO asked before signing off, "Do we have to do this?"

According to Cooke, there was a lot of discussion about the cost premium for green power, especially since it was not a budgeted item. "Even with the vast financial resources of a company like Toyota, it spends its money one yen at a time." Cooke says that they repeatedly argued to senior management that energy conservation could save enough money to cover the increased cost of the green energy.

Public Agencies

Public (government) agencies buying green power include branches of federal (national) governments, local or regional governments, and quasi-public agencies (such as railways and airports). Examples of such purchasers are shown in **Table 7-4**.

Table 7-4: Examples of public/government agencies buying green power

Association of Bay Area Governments (U.S.)
Authorities Buying Council (Scotland, UK)
City of Boulder (U.S.)
London Borough of Lewisham
Los Angeles International Airport
Ministry of Defense (Netherlands)
Nederlandse Spoorwegen (Dutch Railways)
Oxfordshire City Council (UK)
Swedish Railways

Why do these organizations buy green? Federal agencies in the U.S. are buying green for several reasons: to comply with an executive order,[26] to carry out the agency's mission, and in response to a government program that gave awards for support of energy efficiency and renewables.[27] For example, in July 1999, a U.S. Environmental Protection Agency's (EPA's) laboratory in California became the first federal (U.S.) building to be powered by 100 percent renewable energy. This precedent-setting event grew out of the EPA's desire to procure electricity in a manner consistent with its organizational mission of protecting the environment.[28]

Some local and regional governments have set CO_2 or sustainability goals and are using green power to help meet these goals. And some large government agencies have been able to negotiate very low prices for their green power, because of their buying power. When the Authorities Buying Consortium in Scotland bought 112 million kWh per year of renewable energy, it saw an overall decrease in its total energy costs due to a combination of tax exemptions and dropping electricity prices overall.[29]

Nonprofits

Notable nonprofits buying green include environmental organizations, for obvious reasons, and a surprising number of churches. Some of these are very large consumers—the Kirchen Baden-Württenberg (Germany) signed up for 3.7 million kWh per year of green power from NaturEnergie AG.[30] Others have negotiated deals involving co-marketing to their parishioners. In 1999, a group of California churches passed a resolution to purchase renewable power as a way to cut CO_2 emissions. They then made a deal with a green provider to receive a US$20 payment to the church for each parishioner who switched to that provider.[31]

A number of universities have bought green power as well. The University of Edinburgh purchased 25 million kWh per year of wind and small hydropower in 2000—enough to cover 40 percent of its total electricity consumption. There were financial and local economic development drivers for this deal: The university rector posed in front of a local wind farm for a picture in the university newspaper, while the university energy manager noted, "Without this purchase of green electricity, the climate change levy would have added 400,000 Euro to our electricity bill."[32] (See chapter 4 for a discussion of the UK's climate change levy.) Students at the University of Colorado (U.S.) voted to, in essence, tax themselves to purchase the entire output from one wind turbine—an effort led by the university's Environmental Center.

Summary

There are numerous examples of all sorts of organizations and individuals buying green power. Analyses of these buyers reveal some trends—most expected, a few less so. It's not surprising that environmental concerns drive many consumers to buy green, but it is surprising that this same driver works for large corporations as well. The increased cost of green energy is of course a major hurdle, but packaging green energy with overall rate reductions or with energy efficiency can minimize cost increases.

A key policy issue is whether the existence of over 2 million green energy buyers signifies that renewable energy should be "left to the market"—that is, if consumers want it they will buy it, and it should not be explicitly promoted by policy. This chapter suggests, however, that this argument is simplistic. Consumer decisions to buy green are strongly influenced by price, and the electricity industry and electricity prices are heavily influenced by policy. It's not accurate to characterize consumer decisions as made in a policy-free environment.

In addition, green buyers are and will likely remain a minority of all electricity buyers. Leaving renewables to green buyers means that renewables will remain a small piece of the electricity pie. If this is a desired outcome, then leaving it to the green market may be appropriate; but if not, then additional policies need to be considered.

Notes for Chapter 7

1. As of October 2002. From www.greenprices.com/eu/index.asp (downloaded 10 May 2003).

2. Author's estimate.

3. For survey methodology and detailed results, see A. Capage and B. Friedman, "Understanding Residential Green Energy Buyers: A Market Research Survey," E Source Green Energy Series GE-7 (May 2001); and B. Friedman, "Market Research Survey II: Finding Green Energy Buyers," E Source Green Energy Series GE-11 (July 2002). See www.esource.com.

4. Capage and B. Friedman, "Understanding Residential Green Energy Buyers," p. 7.

5. E. Rogers, *Diffusion of Innovations* (New York: The Free Press, 1992).

6. P. Komor, "Making Green Electricity Programs Work: The Experts Speak Out," E Source Green Energy Series GE-5 (September 2000). See www.esource.com.

7. P. Komor, "Making Green Electricity Programs Work."

8. R. Wustenhagen, "Green Power Marketing in Europe—An Overview," presented at the First European Conference on Green Power Marketing, St. Moritz, Switzerland (28 June 2001).

9. Capage and B. Friedman, "Understanding Residential Green Energy Buyers," p. 19.

10. Capage and B. Friedman, "Understanding Residential Green Energy Buyers," p. 19.

11. P. Komor, "Making Green Electricity Programs Work."

12. P. Komor, "Making Green Electricity Programs Work."

13. P. Komor, "Making Green Electricity Programs Work."

14. P. Komor, "Making Green Electricity Programs Work."

15. P. Komor, "Making Green Electricity Programs Work."

16. P. Komor, "Pricing Green Energy," E Source Green Energy Series GE-2 (December 1999). See www.esource.com.

17. P. Komor, "It's Not Easy Being Green: Obtaining Renewable Electricity for Resale," E Source Green Energy Series GE-1 (September 1999), p. 2. See www.esource.com.

18. Capage and C. Hurley, "Making the Big Sale: Selling Green Energy to Large Energy Users," E Source Green Energy Series GE-3 (April 2000), p. 2. See www.esource.com. Also www.greenprices.com.

19. E. Holt et al., "Understanding Non-Residential Demand for Green Power," Report for the National Wind Coordinating Committee (January 2001), p. 2.

20. E. Holt et al., "Understanding Non-Residential Demand for Green Power," p. 27.

21. E. Holt et al., "Understanding Non-Residential Demand for Green Power," p. 33.

22. Capage and C. Hurley, "Making the Big Sale," p. 6.

23. E. Holt et al., "Understanding Non-Residential Demand for Green Power," p. 21.

24. E. Holt et al., "Understanding Non-Residential Demand for Green Power," p. 23.

25. Capage and C. Hurley, "Making the Big Sale," pp. 19–20.

26. An executive order is a requirement instituted by the President of the United States.

27. W. Golove, M. Bollinger, and R. Wiser, "Purchasing Renewable Energy: A Guidebook for Federal Agencies," LBNL-46766 from http://eetd.lbl.gov/ EA/EMP (August 2000), p. 7.

28. This case study is based on W. Golove et al., "Purchasing Renewable Energy," p. 1.

29. ScottishPower news release, "abc and ScottishPower in UK's Biggest Ever Green Energy Deal" (26 March 2001).

30. From www.greenprices.com/de (downloaded 24 April 2001).

31. From www.eren.doe.gov/greenpower/mkt_customer.html#church (downloaded 3 April 2001).

32. As quoted in "Information Note Greenprices" (24 September 2001), from www.greenprices.com (downloaded 21 March 2002).

8

Feed-in Laws: Crude, Effective, Outdated but Not Dead

This chapter looks at "feed-in laws," an anachronistic—yet enduring—tool for promoting renewable electricity. These laws require utilities to buy electricity from renewable generators at high prices, typically at or near the full retail price of electricity. Because these rates are both high and guaranteed by law, investors see little risk in lending for new renewables construction. As a result, investment capital is cheap and easy, and considerable new renewable generation results.

Feed-in laws are in direct opposition to the growing role of competitive markets and pricing—which are replacing regulation and quotas—in the electricity industry. The rates paid to wind generators are seen by many to be unsustainably high, and spreading electricity market liberalization is making it much harder to find a "utility" on which to pin these costs. Yet the past success of feed-in laws means that they have developed a politically powerful constituency that has resisted the phasing out of this heavy-handed policy tool. Additionally, a surprising ruling in 2001 from the European Court of Justice means that the end of feed-in laws is farther away than expected.

Conclusions on the Feed-in Law Approach

Feed-in laws are best summarized as "effective but not efficient" (see **Table 8-1**). They do result in considerable new renewable capacity, but at a high price. They are at best a short-term approach, useful at building up an industry from a low level. Care must be taken, however, that they are limited in both scope and duration, for any subsidy creates its own constituency. In Germany's case, this constituency played a key role in ensuring that the subsidy was extended beyond its economic justification.

Table 8-1: Pros and cons of the feed-in law approach

Pros	Cons
Very effective at getting lots of new renewable generation installed	Reduced incentive for cost reduction
	No direct competition between suppliers
Not a direct general-revenue tax	Price paid reflects outcome of a political process; does not reflect actual costs
Can be very simple	
Costs paid by ratepayers, not general public	Not a "market mechanism"—inconsistent with overall EU direction
Low revenue uncertainty means low-cost capital	Can create "stranded costs" at a further point in restructuring
Low direct cost to government	Can result in "excessive" profits for producers
Little bureaucratic overhead	Sets up a dependent and powerful constituency

The market response to feed-in laws in both Denmark and Germany shed light on the behavior of the capital markets. Renewable energy is, by its nature, more capital-intensive per kilowatt (kW) of generating capacity than fossil-fired electricity generation. (Renewables have zero fuel costs, of course, so they can still be economically competitive.) This capital intensity often means that the availability of capital limits new renewables construction.

Feed-in laws work largely because the capital markets are happy to lend against the security of a guaranteed-by-law future revenue stream. This is not a realistic feature of a market-based economy. However, it does suggest that innovative ways of reducing the perceived uncertainty of future revenues—for example, by long-term contracts with electricity users or by bounding the price of green certificates—can ensure that future renewables will appear attractive to the capital markets.

The U.S.: PURPA

The first large-scale use of the feed-in law approach was the U.S. Public Utilities Regulatory Policies Act (PURPA) of 1978. This act was passed as part of broader federal energy legislation intended to reduce dependence on oil imports, increase energy efficiency, and increase the use of renewables. PURPA essentially required utilities to purchase electricity from small renewable-based generators.

Utilities were required to pay these renewable generators the "avoided cost"—that is, the price the utility would have had to pay to another producer or to build their own capacity, if not for the renewable generator. Although this sounds reasonable, a combination of politics and fossil fuel price volatility resulted in avoided costs being set very high. As discussed below, the unsurprising result was copious amounts of new and expensive renewable capacity.

PURPA left the actual setting of avoided cost to the state utility regulators. In some states, notably California and New York, regulators and the public were both very interested in efficiency and renewables, and less than kindly disposed to utilities due to ongoing problems with nuclear power plants. Oil prices were rising rapidly at that time as well and were predicted by some analysts to reach US$100 per barrel by 1998. (The actual price in 1998 was about US$12 per barrel.[1]) Regulators used these price projections as the basis for defining future avoided costs. In California in the 1980s, utilities were typically paying renewable generators state-defined avoided costs of 10 or more cents per kilowatt-hour (kWh).[2] Furthermore, many states required utilities to sign long-term (ten- to twenty-year) contracts with the renewable generators.

The data on post-PURPA new renewable construction are clear and unsurprising: PURPA led to significant new renewables capacity, but at high prices. The security of high prices guaranteed by long-term contracts with public utilities created powerful incentives for investors and developers to invest in new renewable capacity. As shown in **Table 8-2**,[3] PURPA resulted in some 12.7 gigawatts (GW) of new renewable capacity, of which two-thirds was biomass. In California alone, nonutility renewable capacity increased from 187 megawatts (MW) in 1980 (pre-PURPA) to 4,139 MW in 1996—a more than twentyfold increase.[4]

Table 8-2: New renewable capacity in the U.S. resulting from PURPA

Fuel source	Capacity (MW)
Biomass	8,219
Geothermal	1,449
Hydroelectric	1,263
Wind	1,373
Solar thermal	340
Photovoltaic	14
Total	12,658

Note: Data are as of December 31, 1998, and are for nonutility qualifying facilities (QFs).

The prices paid for this new, PURPA-driven renewable capacity were quite high; the average price in 1995 was 9.05 US cents per kWh. This can be compared to the average price of wholesale electricity at that time, which was 3.53 US cents per kWh.[5] In other words, utilities were paying two to three times more for renewable-based electricity than they were for regular (mostly fossil-based) electricity.

It's important to recognize that these prices reflect the PURPA-based contracts that existed at that time and do not necessarily reflect the true costs of renewable generation. This distinction is not generally acknowledged, however, and PURPA deserves much of the blame for renewables' reputation for high prices, which is pervasive to this day.

PURPA's role and influence has faded over time. When many of the original PURPA contracts came up for renewal in the 1990s, the energy situation had changed. Natural gas prices were relatively low and stable, and natural gas–fired generation was seen as the appropriate measure for determining avoided cost. Utilities and regulators were unwilling to sign long-term contracts at high buyback prices. The State of Idaho, for example, changed its PURPA rules from requiring utilities to sign twenty-year contracts with certain renewable generators to requiring five-year contracts. As a result of these changes, the 1990s saw a marked slowdown in PURPA-based new renewable capacity.

Much of the proposed national restructuring legislation in the U.S. contains provisions to repeal PURPA. Many view PURPA as well-intentioned but outdated in an increasingly market-driven electricity system. As one pro-market advocate argued, "PURPA has saddled utilities with substantial contractual obligations for electricity supply that are not cost-effective, even in today's market, and that clearly will not be viable in a competitive market place. These noncompetitive PURPA contracts are a significant percentage of utilities' stranded costs and an impediment to the development of competitive electricity markets."[6] Renewable generators don't share this view, of course, but the arguments on both sides are muted as state-by-state restructuring, lower avoided costs, and shorter contract terms have all reduced PURPA's importance.

Denmark's Wind Buyback Rules

The oil shock of the 1970s hit Denmark especially hard, as at that time it was using oil for the bulk of its energy needs. The economic damage caused by the run-up in oil prices convinced the Danish government that it needed to look elsewhere for energy, and it then began a vigorous program of wind energy development. A shrewd combination of technology R&D, financial incentives, and thoughtful land-use planning led to a robust and successful wind market. Denmark is a world leader in the wind industry, with 2.9 GW of installed wind capacity as of December 2002 (fourth, behind Germany, Spain, and the U.S.),[7] and with by far the largest per-capita wind generation in the world. Denmark's wind technologies and production companies can now be found worldwide.

Among the many subsidies and incentives used to promote wind energy in Denmark, the feed-in law stands out. This simple law required utilities to purchase wind-generated electricity at a rate that equaled 85 percent of the utilities' production and distribution costs, which worked out to a payment that averaged about 3.8 US cents per kWh.[8] In addition, privately owned turbine owners received a direct subsidy of 3.2 US cents per kWh.[9] These two payments add up to a hefty 7.0 US cents per kWh. There were also some tax exemptions for privately owned wind turbines.[10] As a result of these various incentives, wind power has been a consistently profitable investment for private investors in Denmark.

Eligibility for the feed-in tariff changed over time, reflecting shifts in policy goals. Until the early 1990s, eligibility for the 85 percent feed-in rate was restricted. For example, individually owned turbines were eligible only if they were smaller than 150 kW. Cooperatively owned turbines were eligible only if they were located in or near the owning cooperative. These size and ownership restrictions were eventually eased.[11]

In contrast with other countries, private and cooperative investors rather than corporate or commercial investors own the bulk of Denmark's wind turbines. This reflects in part Denmark's cultural focus on cooperatives but is largely a result of an explicit national policy to promote private investment in wind.[12] It's no surprise then that 100,000 Danes (2 percent of the Danish population) hold a stake in a wind investment.[13]

The active role played by private and cooperative investors in the Danish wind market has done much to ameliorate the planning problems faced by wind in much of Western Europe. Whereas local residents in the UK, for example, often view wind turbines as an unwelcome intrusion that benefits utilities or other institutional investors, Danes see direct financial benefits from wind investments and thus are more open to having them located nearby.

The widely recognized success of Denmark's approach (using the admittedly narrow definition of success as amount of new renewables built) makes it all the more interesting that Denmark in the late 1990s moved to largely abandon the feed-in approach in favor of the then-trendy yet unproven green certificate approach. There were several drivers behind this change:

- The expectation that the electricity industry would become increasingly liberalized and market-driven, making the feed-in law approach inappropriate.

- The belief that, sooner or later, EU rules would prohibit the feed-in approach (a belief disproven by the EU Court's ruling on the German feed-in law, discussed below).

- Growing complaints from large industrial energy users, who pay for the feed-in requirement via higher electricity rates.

- Concern from government budget officials who saw a large and growing expense from the 3.2-US-cents-per-kWh subsidy for wind producers.

These forces led to legislation that phased out many of the various financial incentives for wind (including the feed-in tariff) and replaced them with a green certificate system. Denmark's commitment to renewable energy has not weakened. The country has committed to generate 20 percent of its electricity from renewables by 2003 (a doubling from 1998 levels).[14] Denmark has also committed to a 20 percent reduction in CO_2 emissions by 2005, relative to 1998 levels. But its methods of achieving these goals have changed.

Under the proposed new system, generators would receive green certificates reflecting their renewable generation. Electricity users would be required to purchase green certificates to cover a fraction of their electricity consumption (the specific amount has not yet been decided). The cost of the green certificates was bounded to be between 1.2 to 2.3 US cents per kWh. The price of the electricity itself would be market-based.

Not surprisingly, the wind industry was none too excited about the changes. As the general manager of the Danish Wind Turbine Manufacturers Association noted, "If the reform proposals are fully implemented, the volume of orders will slow to a trickle...the problem is that no bank will give a ten-year loan to a windmill if it only knows the price conditions for four years of the wind turbine's operating life."[15] But this view proved too pessimistic. Due to the generosity of the transition rules, there was a rush on wind in 1999 and 2000. As a result, Denmark expected to reach 27 percent renewables by 2003—far in excess of the 20 percent goal.[16] In 2001, the proposed green certificate system was placed on hold, due largely to opposition from the renewables industry (see chapter 11 for more information on green certificates).

Germany's Feed-In Law

Germany, like Denmark, has a generous feed-in law that has resulted in prodigious installations of new renewable capacity. It is a controversial law, however, because its very success has meant that some utilities in windy regions have ended up with more wind capacity than they would like. Recent modifications and a supportive court ruling suggest that Germany, unlike Denmark, will likely keep its feed-in law on the books, despite its nonmarket nature.

Germany's electricity feed law of 1991 was startlingly short, simple, and effective at promoting wind turbine installations. Key points of the law were as follows

- Utilities were required to purchase electricity from wind and solar generators at a rate equal to 90 percent of the utilities' average revenue.

- For hydro-, landfill gas–, sewage gas–, and biomass-fueled generators smaller than 500 kW, the rate was 80 percent of average revenue, dropping to 65 percent for facilities over 500 kW.

- If a utility ended up purchasing more than 5 percent of its electricity at these required rates, it would be reimbursed from its upstream system operator. (This amendment to the original law was added in 1998.)

As electricity prices in Germany were quite high, this law made wind turbines a nice investment. Wind generators, for example, received 9.7 US cents per kWh from utilities in 1998 (**Table 8-3**).[18] These rates were stable as well, fluctuating by only a few tenths of a cent from 1991 to 1998 (**Table 8-4**).[19]

Table 8-3: Germany's electricity feed law buyback rates (1998)

Technology	Percentage of average revenue	Corresponding rate in US cents/kWh
Wind, solar	90	9.7
Small hydro, biomass	80	8.6
Large hydro, biomass	65	7.0

Notes: Based on an electricity rate of 18.66 pfennigs/kWh and a conversion rate of DM1.734 = US$1.

Table 8-4: Wind and solar buyback rates in Germany

Year	Price paid to wind and solar generators (US cents/kWh)
1991	9.6
1992	9.5
1993	9.6
1994	9.8
1995	10.0
1996	9.9
1997	9.9
1998	9.7

Notes: Conversion rate of DM1.734 = US$1 assumed.

Due largely to the feed-in law,[20] Germany's installed wind capacity jumped from a few megawatts in 1990 to over 12,000 MW by the end of 2002—the largest capacity of any country.[21]

Utilities were, not surprisingly, unhappy with the feed-in law. Their arguments against it included these:

- It was an unnecessary subsidy for older hydropower plants.

- It led to the installation of uneconomic wind turbines.

- It was expensive. According to utility estimates, in 1997, German utilities paid out US$235 million more for renewable generation than could be economically justified.[22]

The utilities lobbied at the federal level to modify the law and, in addition, raised a legal challenge to the law with the European Court of Justice, arguing that it was an inappropriate and unjustified subsidy in violation of EU regulations.

The utilities' dislike, however, was balanced by the renewable industry's wild enthusiasm for the law. Wind developers had a high and legally guaranteed price for their output and thus were able to secure inexpensive financing for new turbine installations. And as the German wind industry grew in the 1990s, so did the political power of the industry, which aligned itself with the German Green party and independent environmental groups. However, other technologies, notably biomass and photovoltaics, received less benefit from the feed-in law, and backers of those technologies lobbied for change.

The combination of high costs, inconsistency with the EU's antisubsidy philosophy, and pressure from utilities and some renewable groups led to passage of the Renewable Energy Law, effective April 2000, which replaced the old feed-in law. The Renewable Energy Law set specific buyback rates by technology (**Table 8-5**)[23] and included provisions to decrease those rates over time. It also provided for a 0.04 US cents per kWh surcharge on all electricity use to fund the buyback.

Table 8-5: Buyback rates in the 2000 German Renewable Energy Law

Technology/source	Buyback rate (US cents/kWh)
Hydropower, landfill gas	6.5
Biomass <500 kW	8.7
Wind	7.7
Photovoltaics <100 kW	43.0

This new law led to a mad rush to install photovoltaics (PVs), as a buyback rate of 43 US cents per kWh was high enough to make even this high-priced technology a profitable investment. In short order, much of the world's PV production was flowing to Germany. This led to some unexpected problems, notably the diversion of PVs from less-profitable rural electrification projects in the developing world to the more-profitable German market.[24] The availability of low-cost loans in Germany for PV projects added further to the profitability of this market.

Early in 2001, the European Court of Justice responded to a case involving the original feed-in law. The law had been challenged by a utility on two grounds: that the mandated payments to renewable generators constituted "state aid," which is not allowed under EU rules, and that, because only German renewable generators were eligible, it violated EU open market rules. The court ruled against the utility and for the law on both counts. The court found that the payments were not state aid, as the funds came not from the government but from the utility. And, somewhat surprisingly, the court opinion indicated that environmental considerations could supercede open market rules.

Although the court's ruling was a response to the original feed-in law, it had implications for the Renewable Energy Law—and for other EU countries' use of the feed-in requirement policy approach. One analyst noted, "[the ruling] will strengthen the hand of (EU) member states that wish to promote renewables through high fixed tariffs rather than a harmonized market-oriented system."[25] In essence, the ruling gave a bit of a reprieve to the feed-in approach, which is philosophically inconsistent with the EU's clear preference for less direct regulation and greater use of market signals and approaches.

In summary, feed-in laws do yield considerable new renewable capacity, but at high prices. And feed-in laws are a poor fit with the growing role of prices and competitive markets in electricity. The U.S. feed-in law is fading while Denmark's is in limbo. The combination of entrenched political influence and a surprising favorable court ruling have proven to be more than a match for the restructuring movement in Germany; however, this is likely a short-term result. The overall trend is clear, and feed-in laws do not have a bright future.

Notes for Chapter 8

1. U.S. Department of Energy, Energy Information Administration, "Renewable Electricity Purchases: History and Recent Developments," *Renewable Energy Annual 1998: Issues and Trends* (March 1999), p. 9.

2. U.S. Department of Energy, "Renewable Electricity Purchases," p. 9.

3. U.S. Department of Energy, Energy Information Administration, "Incentives, Mandates, and Government Programs for Promoting Renewable Energy," *Renewable Energy 2000: Issues and Trends* (February 2001).

4. U.S. Department of Energy, "Incentives, Mandates, and Government Programs," p. 12.

5. U.S. Department of Energy, "Incentives, Mandates, and Government Programs," p. 13.

6. J. Eisenbach and T. Lenard, "How to Recognize a Regulatory Wolf in Free Market Clothing," *Progress in Point* (Progress and Freedom Foundation, July 1999), further information at www.pff.org.

7. American Wind Energy Association (AWEA), "Global Wind Energy Market Report" (February 2003), from www.awea.org (downloaded 20 May 2003).

8. In 1998. Danish Energy Agency, "Wind Power in Denmark: Technology, Policies and Results" (September 1999), p. 12, from www.ens.dk/uk/publica.html (downloaded 21 May 2001). Conversion rate of US$1 to 8.5 Danish Kroner (DKK) assumed.

9. Danish Energy Agency, "Wind Power in Denmark," p. 12. See also Paul Gipe, *Wind Energy Comes of Age* (Wiley, 1995), p. 60.

10. Danish Energy Agency, "Wind Power in Denmark," p. 11.

11. Paul Gipe, *Wind Energy Comes of Age*, pp. 60–61.

12. Danish Energy Agency, "Wind Power in Denmark," p. 8.

13. Financial Times, "Denmark Wind: Government to Meet Half the Country's Energy Needs Through Offshore Wind," *European Energy Report*, no. 533 (4 June 1999).

14. Danish Energy Agency, "Annexes to the Report on the Green Certificate Market" (December 1999), from www.ens.dk/uk/publica.html (downloaded 21 May 2001), p. 45.

15. As quoted in Financial Times, "Danish Wind Turbine Makers Fear Reform Plan," *Renewable Energy Report*, no. 1 (26 March 1999).

16. Niels Ladefoged, Danish Energy Agency, personal communication (31 May 2001).

17. Original legislation available in English at www.loy-energie.de/gesetze/feed-law.htm (downloaded 24 May 2001).

18. International Energy Agency, "Renewable Energy Policy in IEA Countries, Volume II: Country Reports (1998)," p. 112, see www.iea.org. European Commission, "Electricity from Renewable Energy Sources and the Internal Electricity Market," Working Paper (not dated), see europa.eu.int.

19. European Commission, "Electricity from Renewable Energy Sources and the Internal Electricity Market."

20. There were additional incentives in place as well that promoted wind, including low-interest loans and capital grants, but the feed-in law was the most influential. See International Energy Agency, "Renewable Energy Policy in IEA Countries," pp. 111–118.

21. American Wind Energy Association (AWEA), "Global Wind Energy Market Report," (February 2003), available from www.awea.org (downloaded 20 May 2003).

22. Financial Times, "Utilities Say State Support for Renewables Should Replace Feed-In Law," *European Energy Report*, no. 517 (16 October 1998). Converted to U.S. dollars at 1.7 DM per US$1—the 1997 conversion rate.

23. P. Maegaard, "Sensational German Renewable Energy Law and Its Innovative Tariff Principles," presented at the Eurosun 2000 Conference, Copen-

hagen, Denmark (20 June 2000). Conversion rate of US$1 = DM2.3 assumed.

24. See "German Incentives More Than Fill PV Order Books—How Will the Industry Respond?" *Renewable Energy World* (May-June 2000).

25. Financial Times, "ECJ Ruling Stills Debate on Feed-in Laws—For Now," *Renewable Energy Report*, no. 26 (April 2001), p. 4.

9

The UK's "Tender" Approach: Price Reduction, but at a Cost

The NFFO (Non-Fossil Fuel Obligation) was the UK government's main tool for promoting renewable electricity in the 1990s, and its successes and failures point to several useful conclusions about how to craft effective renewable policies.

The NFFO had its origins in the Electricity Act of 1989. That legislation established the government's authority to require electricity companies to purchase non-fossil fuel–based electricity and also established an electricity tax to be used to cover the above-market costs of this non-fossil generation. These sections of the legislation were originally intended to provide a subsidy for nuclear power (hence the name "non-fossil"), which was then seen as vulnerable due to the impending competition in electricity supply. (See chapter 3 for a discussion of competition.) The concept, however, was extended to renewables in 1990.

How the NFFO Worked

The NFFO was essentially a supply-side subsidy combined with a bidding process. The supply-side subsidy came from an electricity tax, known as the fossil-fuel levy. This was initially set at about 10 percent of electricity prices, with more than 90 percent of the tax revenue going to nuclear power. Over time, as the nuclear subsidy was reduced, the tax rate fell to less than 1 percent of electricity prices.

The portion of the tax revenue from the fossil-fuel levy set aside for renewables was used to cover the difference between the wholesale market price of electricity (which was determined largely by fossil-fired generation) and the cost of renewable electricity. For example, if the prevailing wholesale market price for regular electricity was 3 US cents per kilowatt-hour (kWh), while renewable electricity's wholesale price was 4 US cents per kWh, the regional electricity companies

(RECs) paid the renewable generator 3 US cents per kWh, and the electricity tax revenues were used to give the renewable generator the difference of 1 US cent per kWh. The RECs were required to buy the renewable electricity, but not to cover its premium costs.

The bidding component of the NFFO came from the process used to determine the price to be paid to the renewable generators. Renewable generators provided a bid—essentially an offer to provide a certain amount of renewable electricity at a certain price—and the government then accepted all bids at or below a cut-off ("strike") price. For each of the five rounds, or bidding cycles, the government set a strike price within a distinct technology "band," or type of renewable generation. For example, for the first-round bidding cycle, the maximum price for wind power was set at 15 US cents per kWh.[1] All offers received by the deadline for wind power at or below that price were accepted, and all that accepted capacity was (in theory) bought by the RECs. To encourage a range of renewable technologies, different types of renewable generation had different maximum prices. For example, for the first round of bidding, the maximum allowable price for landfill gas was set at 9 US cents per kWh[2]—quite a bit lower than wind's 15 US cents per kWh.

The Results

There were five bidding rounds from 1990 to 1998, and changes in technologies, prices, and capacity built over this time reveal how this policy affected the market.

Allowed Technologies

For this program, as for most renewable programs, there was some disagreement over just what types of renewable technologies should be eligible for the subsidy. Wind, hydropower, and landfill gas were eligible for all five rounds, while other technologies such as sewage gas and municipal waste burning were only eligible for some rounds (**Table 9-1**).[3] Shifts in which technologies were eligible were due in part to political factors: Every technology has its advocates, and most saw NFFO as a beneficial subsidy and therefore pushed to be included.

Table 9-1: Eligibility of selected renewable technologies for the NFFO program

Technology	NFFO1 (1990)	NFFO2 (1991)	NFFO3 (1994)	NFFO4 (1997)	NFFO5 (1998)
Wind	X	X	X	X	X
Hydropower	X	X	X	X	X
Landfill gas	X	X	X	X	X
Sewage gas	X	X			
Gasified biomass			X	X	
Farm wastes				X	
Municipal waste with CHP[a]				X	X

Note: (a) CHP = Combined heat and power, also known as cogeneration.

Prices

Prices were set separately for each technology, which allowed the NFFO to promote several different renewable technologies, not just the cheapest. This led, predictably, to charges of what is sometimes called "industrial policy," or more simply "the government picking winners." This was considered by many to be inconsistent with a promarket policy and therefore bad. As this illustrates, probably the most difficult part of setting renewables policy (or policy in general) is accommodating different and at times conflicting policy objectives. Was the objective of NFFO to promote wind power, to promote cost-effective renewables, to drive costs down, to get the most renewable generation for the least amount of money, or to build a domestic renewables industry? The answer is "yes"—that is, each goal was desirable and had its constituency. Setting different prices for each technology was an attempt to provide some subsidy to each technology, even if it was less cost-effective than another.

For the first round, the strike prices—that is, the maximum allowable prices for each technology—were quite high. This was, to some extent, an explicit market promotion strategy: The government wished to get the renewable market underway with a relatively generous initial boost. After that, however, prices dropped dramatically—wind power, for example, dropped from 15 US cents per kWh in the first round to 4.3 US cents in the final round. Other technologies also saw large price drops (**Table 9-2**).[4] Many consider these remarkable price reductions as the NFFO's greatest success.

Table 9-2: Prices dropped sharply in later bidding rounds

Technology	NFFO1	NFFO2	NFFO3	NFFO4	NFFO5
Wind	15	16.5	6.6	5.3	4.3
Hydropower	11.3	9.5	6.7	6.4	6.1
Landfill gas	9.6	8.6	5.6	4.5	4.1
Sewage gas	9.0	8.9			
Municipal waste with CHP[a]				4.8	3.9

Notes: All prices are in US cents per kWh.
(a) CHP = Combined heat and power, also known as cogeneration.

A closer look at the cost data reveals that:

- Wind prices seen by the NFFO were about half those of wind in other countries.[5]

- Technologically mature technologies, such as hydropower, saw an almost 50 percent cost reduction from the first round in 1990 to the fifth round in 1998.

These two observations suggest that the dramatic cost reductions seen in just eight years were not only due to technological improvements, but also to improvements in the myriad factors that determine final market costs. These include managerial efficiency, cost of capital (which is, in turn, determined by perceived risk), operating efficiency, and profit margins.

Getting It Built

The success of the NFFO at driving down prices was unfortunately not matched in its translation of successful bids into actual projects. The first round saw relatively high execution rates (defined as the percentage of successful bids that ended up as actual generation), but this was because many of those projects involved modifications to existing power plants. Further rounds saw disappointingly low execution rates—for example, only about one-third of the waste projects whose bids were accepted in round 3 ended up actually generating electricity (**Table 9-3**).[6]

**Table 9-3: Renewable capacity contracted and built
for NFFO round 3**

Technology	MW contracted	MW built	Percentage
Hydropower	15	12	80
Municipal and industrial waste	242	77	32
Small wind	20	10	50
Large wind	146	35	24

There were two main reasons for these low execution rates: unrealistically low bidding by bidders and unexpected difficulties in local permitting. In the rush to ensure that their proposals got in on time and under the strike price, it appears that some developers made unrealistic assumptions about their own costs. The NFFO also allowed about a five-year window from acceptance of a bid to actual generation. Developers may have been banking on hoped-for price reductions that did not materialize. As there was no penalty for failure to deliver on an accepted bid, developers had little to lose from submitting an overly optimistic bid.

Local permitting problems, especially for wind power, were legion. The NFFO suffered from this as well—and as developers were awarded NFFO contracts at the beginning of the planning and permitting process, they had a reduced incentive to ensure that their proposal was for a site for which they would be successful in getting the needed local permits. They did not get paid if the project was not built, of course, so they clearly had some incentive to get it right, but early in the NFFO process, few anticipated how hard it would be to get through the local permitting process. The lack of a stick to complement the NFFO subsidy carrot turned out to be a serious flaw.

What Worked

The greatest overall success of the NFFO was price reduction. The sharp drops in delivered per-kilowatt-hour costs (see Table 9-2) suggest that the NFFO was successful at getting renewable generation out of the one-off, technology-driven mindset and into the market-based, competitive product world. Although correlation is not necessarily causality—that is, the fact that prices dropped while NFFO was underway does not necessarily mean that NFFO *caused* the price drop—the dramatic price drops seen over the successive NFFO rounds, coupled with NFFO's price-based structure, make it likely that NFFO did play a large role in driving the price drop.

A subtler but still handy benefit of the NFFO was that its staged implementation—in a series of rounds rather than all at once—allowed for fine-tuning of procedures, prices, allowed technologies, and other details by round.

What Didn't Work

This discussion of problems with the NFFO may look rather daunting, but most of these "problems" are not really shortcomings of the NFFO itself, but rather disagreements over what goals the NFFO should have been trying to accomplish. They are generally problems only in the eyes of those whose goals weren't addressed.

Fundamental to any discussion of the NFFO's problems are the inherent contradictions of pursuing a pro-market policy. Almost by definition, a policy is a market intervention intended to accomplish some goal—a goal that presumably would not be met if the policy didn't exist. Many of the NFFO's criticisms reveal this tension.

The "picking winners" argument—that using technology bands meant that some technologies were favored over others—is a classic argument. It's true, of course, that the NFFO did provide a higher price for wind than for landfill gas, for example. But it's also true that the NFFO "picked" renewables over other electricity generating technologies. NFFO used the technology bands in order to encourage several different renewable technologies, rather than just the least expensive. It's this policy goal, rather than the NFFO itself, that is of contention.

Similarly, the NFFO did nothing to further the market readiness of noncommercial technologies, such as photovoltaics (PVs). (Note that this is the flip side of the "picking winners" argument.) Although the NFFO could have been structured with bands for relatively expensive technologies such as PVs, it was not. Here again, the disagreement is over policy goals: Should the NFFO be directed at only the cheapest renewable technology, at a range of close-to-market technologies, or at all renewable technologies? The NFFO took a middle ground and thus was criticized by those at both ends.

Along these same lines, some were disappointed that the NFFO did little to encourage a domestic (UK) renewable technology supply industry. Developers selected the least-expensive, most market-ready technology; in the case of wind, this usually meant a Danish manufacturer. Similarly, some looked to the NFFO to encourage new, small companies to enter the renewables market, but here again NFFO's focus on cost gave an advantage to large, established companies with ready access to lower-cost capital.

As noted earlier, NFFO's greatest procedural problem (that is, as distinguished from those related to its choice of goals) was its poor execution rates—that is, its failure to deliver actual, in-the-ground renewable capacity instead of just signed contracts. A ready fix for this—unfortunately at the expense of contract complexity—would be to incorporate a stick of some sort, such as a bond upon contract signing that is returned when actual generation starts.

Epilogue

In the late 1990s, the UK government undertook a series of energy policy reviews. Among the many policy changes that resulted from these reviews was the replacement of the NFFO with a renewable "obligation" (see chapter 10). It's not accurate, however, to characterize the NFFO as a failure that was eliminated for nonperformance. It certainly had its problems, but it may have been the right tool for its time—when many renewable technologies were technically mature but not market-ready. By establishing a guaranteed price for renewable-sourced electricity, it reduced risk to the level where renewable project developers were able to access reasonably priced capital. By about 2000, several renewable technologies, notably wind and some forms of biomass, were close to cost-competitive with fossil fuels—due in part to the NFFO. And an interesting twist on NFFO was under consideration: a reverse bid, in effect, in which the NFFO-supported renewable electricity would be auctioned off.

Notes for Chapter 9

1. C. Mitchell, "The England and Wales Non-Fossil Fuel Obligation: History and Lessons," *Annual Review of Energy* (2000), p. 292.

2. C. Mitchell, "England and Wales Non-Fossil Fuel Obligation," p. 292.

3. C. Mitchell, "England and Wales Non-Fossil Fuel Obligation," p. 291.

4. C. Mitchell, "England and Wales Non-Fossil Fuel Obligation," p. 292.

5. C. Mitchell, "England and Wales Non-Fossil Fuel Obligation," p. 290.

6. C. Mitchell, "England and Wales Non-Fossil Fuel Obligation," p. 296.

10

The Renewable Portfolio Standard

A growing number of countries are using a very simple approach to get new renewable capacity built: They are setting a mandatory goal for renewables content and letting the market find the least expensive way to get there. Although the names (renewable portfolio standard, renewable obligation, and quota) and details vary, the fundamental idea is simple, clear, and—in most cases—quite effective. This approach has wide political support. It gets the nod of approval from both the free market supporters, who like its basis in price, and the renewables advocates who like the certainty of an explicit goal. But it has had differing effects on different renewable technologies, and this is both its strength and its weakness. In its purest form it differentiates only on price, which is sometimes not the result that policy makers—and various advocates—want. As with all policies, careful implementation is key.

What Is the Renewable Portfolio Standard, and How Does It Work?

Most countries have goals for renewables, but the existence of such a goal alone does not make it a renewable portfolio standard (RPS). (We'll use the U.S. term *RPS*. In the UK, it is known as an *obligation,* while other European Union [EU] countries generally use the term *quota*.). RPSs are distinguished by:

- Assigning responsibility for meeting the goal to a specific actor—such as electricity users, suppliers, or generators.

- Having a substantive penalty for failing to meet the goal.

156

Most RPSs are based on an explicit annual goal for renewable generation, which can be defined as a percentage of total electricity generation (for example, 10 percent of generation must come from renewable generation by 2010) or as new capacity (for example, 200 megawatts [MW] of new renewables must be added by 2005). These goals usually increase annually. Responsibility for building or buying the required renewable electricity is apportioned in some way—for example, most RPSs require electricity retailers to provide a certain percentage of the electricity they sell as renewable. Most RPSs allow trading, making this policy option closely linked with the idea of green certificates (see chapter 11). Some RPS programs allow for "buying out"—essentially setting a cap on costs—while others have a steep penalty for noncompliance.

Pros and Cons

The RPS is emerging as one of the more popular options for promoting renewable electricity generation. It works better in some situations than others, and its success strongly depends on the many details of its implementation. Like all policy options, it can't be all things to all people—so the more clarity and agreement one can achieve on the policy goals up front, the greater the chances of success.

Its principal advantages include:

- *A specified amount of renewable generation is ensured.* A successful RPS will yield the required amount of renewable generation—unlike other policy options such as financial incentives, whose impacts are often hard to predict.

- *Administrative and bureaucratic costs are low.* Once the goal is set, the only significant government role is monitoring and enforcement. This is typically done through reporting requirements—an annoying but not overwhelming paperwork burden.

- *It combines well with trading of renewable credits* (see chapter 11). In fact, most RPSs allow such trading.

- *Price pressure is maintained.* There is no guaranteed price, so (in theory) renewable generators will feel continual market pressure to reduce prices.

- *Risk is reduced, although not eliminated.* An RPS establishes a guaranteed demand for renewable electricity but does not guarantee demand for any specific generator. So overall market risk is reduced but not eliminated—meaning, for example, that reasonably priced capital will be made available.

- *It is simple.* Although there are some interesting subtleties, overall this is one of the simplest and most transparent policy options. This makes it politically attractive.

The RPS approach does have some problems as well:

- *It does not deal well with differing costs across technologies.* A typical RPS has a list of qualifying technologies—but higher-priced ones, such as photovoltaics, will not benefit from the RPS, as they will be unable to compete financially.

- *The renewables goal is set politically and is not price or performance based.* An electricity system uses a mix of generation that is largely determined by price, reliability, and other technical and economic factors. The RPS goal, in contrast, is the outcome of a political process.

- *It is not fundamentally a market mechanism.* The RPS is a regulation and thus imposes some costs.

Lessons Learned

A number of EU countries and U.S. states already have RPS-like requirements in place. Although many of these have been operating for only a short time, they have already provided lessons about the RPS approach.[1] These lessons include the following:

- *Clarify policy goals.* Is the goal to reduce carbon, to build a domestic renewables industry, to promote fuel diversity? The answer should steer the list of included technologies and other details of the RPS implementation.

- *Set the goal correctly.* The ideal RPS goal will increase the use of renewables to a level higher than what would otherwise occur yet will not unduly increase electricity costs or excessively goad political opponents. Setting the goal needs to be a political process, not an analytical one, and should if possible be the result of consensus.

- *Include technologies carefully.* There is not universal agreement on just what qualifies as "renewable." Every technology has its advocates, and all will want the RPS to benefit them. Including municipal waste, for example, will engender political support from waste plant developers but will likely result in opposition from environmental groups. The list of included technologies should reflect the policy goal.

- *Expect dissent from higher-priced technology stakeholders.* Photovoltaics, for example, will not benefit from an RPS unless it contains special provisions to accommodate this more expensive technology.

- *Allow for trading.* This will reduce costs.

The UK's Renewable Obligation[2]

The Non-Fossil Fuel Obligation (NFFO; see chapter 9) was a mixed bag for the UK. Although it did drive down costs, it was complicated, bureaucratic, and failed to deliver on much of its contracted capacity. It was replaced by the Utilities Act of 2000, which, through its Renewable Obligation, took a very different approach to encouraging renewables:

- It set explicit goals for renewable electricity generation.

- It established that this "obligation" would be the explicit responsibility of electricity retailers.

- It established a green certificate system to allow for trading of renewable generation.

- It set up a buyout price that allowed suppliers to, in effect, buy their way out of the obligation.

Subsequent government reports and other sources laid out how the obligation would actually work. These details, and how they've changed to reflect political realities and a better sense of how the market responds to various incentives, provide insight into how policies and markets interact.

Fuels and Technologies Included

Renewable technologies that can be used to meet the Renewable Obligation are landfill gas; sewage gas; energy from non-fossil waste; small (20-MW or less) hydropower; onshore and offshore wind power; most forms of biomass; and geothermal-, tidal-, wave-, and photovoltaic-based power. As first proposed, large-scale hydropower and energy from municipal solid waste were explicitly excluded, because according to a government report they "are already commercially viable, well established in the market, and can compete with electricity from fossil fuels."[3] This quote nicely illustrates the multiple and sometimes conflicting policy goals that renewable policies must contend with. If the goal of the

Renewable Obligation was simply to meet the 10 percent goal at the lowest cost, it would make no sense to exclude those technologies that are the lowest cost (which would likely mean large hydropower). But the goals are not that simple; they also include promoting less-competitive technologies. In this case, the UK first took the route of excluding large hydropower and municipal solid waste—both of which were controversial due to their environmental impacts, as well as inexpensive. Interestingly, in 2002 new large hydropower was put back on the list of qualifying technologies, due to "concern...expressed by the (hydropower) industry...that some potential new developments could not proceed without support."[4]

How Different Fuels and Technologies Are Treated

Renewable technologies, however defined, vary considerably in cost, environmental impact, lead time for construction, and other attributes. Should a renewables policy treat different technologies differently? This is a difficult and controversial question, and one that is at the heart of what is called the "banding" question. The UK's NFFO set up different "bands" for different technologies, in order to allow those that were more costly to still receive some benefits from the program (see chapter 9). For the Renewable Obligation, however, the UK has taken a path that is somewhat less discriminating. The UK chose not to use a banded approach—each renewable kilowatt-hour (kWh), regardless of source, counts the same. However, the first round of proposed implementation rules included a supplemental program of capital grants to be used to further subsidize offshore wind and energy crops.

A report by the UK's Department of Trade and Industry (DTI) reveals some of the thinking that led to that approach. Although the language is a bit self-contradictory, it does show the tension faced among at-times incompatible goals when setting renewables policies:

> We believe that a banded obligation would segment the market unnecessarily, and would lead to Government dictating the relative importance of each technology. We also feel that it is no longer Government's job to pick winners or to introduce artificial distortions in the marketplace....In recognition of the competitive challenge this poses for the industry, we are proposing additional support for an initial round of both offshore wind and energy crops projects.[5]

There is a political element here as well. Regardless of the policy chosen, some will benefit more from it than others. In a democratic society, ensuring policy success requires obtaining the right amount of support—ideally from as many different groups as possible. This often means policies must be modified to reflect political reality.

Industry reactions to the proposed capital grants scheme were generally negative. The biomass industry pointed out that such a scheme put them at a disadvantage, as they have lower capital costs and higher fuel costs than wind. This means that a scheme that, for example, covered 30 percent of capital costs would mean less financial support, per unit of output, for biomass than it would for wind. As the Renewable Obligation was being implemented, there were serious proposals for what was called a "double ROC (renewable energy certificate)"—certain technologies such as biomass and offshore wind would get two green certificates per kilowatt-hour, while all other included renewable technologies would get only one. As of 2002, however, it appeared that the double ROC was dead. A government document noted, "(banding) would be contrary to the market-led basis of the Obligation and would remove the essential ingredient of competition between renewable energy technologies."[6]

The Schedule

The obligation was intended to officially commence in October 2001, but, due in part to the spring 2001 election, was delayed and did not fully begin until spring 2002. The required renewable percentage increases annually, starting at 3.0 percent of sales in 2003 up to 10.4 percent by 2011 (see **Table 10-1**).[7]

Table 10-1: Required renewable contributions in the UK's Renewable Obligation

Year	Required renewables as a percentage of total UK electricity sales
2003	3.0
2004	4.3
2005	4.9
2006	5.5
2007	6.7
2008	7.9
2009	9.1
2010	9.7
2011	10.4

Although explicit renewables goals are set through 2011, the UK's "preliminary consultation" (a government report that lays out a detailed proposal for how the policy will be implemented) took pains to make clear that the program was intended to last for much longer than that. The consultation stated, "The obligation will remain in force until 2026 and will provide a guaranteed market for electricity generated from renewable sources until that date. Such long-term commitment is an unusual step for Government but highlights the importance attached to this particular energy issue."

This was likely an effort to reduce the perceived political risk—the risk that lenders see in projects whose future revenues are dependent on policy not changing. It's a bit misleading, however, as policies are always in flux, reflecting both swings in politics and fine-tuning that takes place as lessons are learned along the way. Promoting the program as long-term and stable has a downside as well: The longer and more entrenched a subsidy of any sort, the more it is perceived by the beneficiaries as a right rather than a short-term boost.

The point is moot if the Renewable Obligation is successful, because it will in effect put itself out of business. If it succeeds in making renewables cost-competitive with fossil fuels, then the renewable percentage will rise without the market pull of a Renewable Obligation.

Buying Out

The UK system puts responsibility for meeting the Renewable Obligation squarely on retailers—those selling electricity to end users. All licensed electricity retailers have to provide evidence to the energy regulatory office Ofgem (Office of Gas and Electricity Markets) that they have met the obligation. They can do so three ways:

- By presenting green certificates that they have obtained for delivery of eligible renewable production to their customers,

- By presenting green certificates that they have purchased independently of the generation that yielded those certificates, or

- By paying the buyout price, proposed at 4.4 US cents per kWh in 2002.

The buyout price is an interesting component to the Renewable Obligation policy. The level of this buyout price is crucial, as it sets an upper bound on the market value of renewables. In other words, suppliers will generally (but not always; see below) pay no more than this for renewable electricity, as it would be cheaper to simply pay the buyout price.

The selection of a specific buyout price essentially involves trading off cost and risk. The intent of a buyout price is to set an upper limit on the cost of renewables (and thus the cost of the entire program). The higher the buyout price, the higher the potential monetary cost of the program (see **Table 10-2**).[8]

Table 10-2: Estimated cost to consumers of the UK's Renewable Obligation program

Buyout price (US cents/kWh)	Cost to consumers (US$ millions)
3.0	710
4.4	890
6.0	1,210

So why not simply choose a lower buyout price and reduce the cost of the program? Because doing so increases the risk that the overall goal will not be met. Imagine, for example, a very low buyout price, of say 0.1 US cent per kWh. At this level, virtually all suppliers would find it less expensive to buy out rather than buy renewables, and thus there would be no market for new renewables.

As always, a political component needs consideration as well. The final cost to consumers is of concern to consumer groups, so one might expect them to argue for a lower buyout price. In the UK's case, however, the principal UK consumer lobby, the National Electricity Consumers Committee, was supportive of the Renewable Obligation and did not object to the 4.4 US cents per kWh buyout price. This group saw renewables as a long-term way to reduce the impacts of rising natural gas prices. The chairman of the Consumers Committee noted, "Over the longer term [the Renewable Obligation] may prove to be not only in the environmental interest but consumer interests as well."[9] The renewables industry generally doesn't like buyout prices, as they set a limit to the prices they can charge. In this case, however, the proposed 4.4 US cents-per-kWh buyout price was an increase from previous proposals, so criticism was minimal.

The revenues from the buyout system are recycled to retailers in proportion to the renewables they provide. In other words, if a retailer provides all the renewable electricity it is required to, and this renewable electricity amounts to 10 percent of all renewable electricity, then that retailer receives 10 percent of all buyout revenues received. This provides an additional incentive to retailers to meet the Renewable Obligation, because, if they fail to do so (and therefore pay the buyout price), they will end up subsidizing their competitors. This may result in the unexpected result of suppliers paying more than the buyout price for renewables, in order to keep their funds out of the hands of their competitors.

Market Marxism?

The degree to which the UK's policy veers away from a pure market approach is well illustrated by a quote from the UK's DTI: "The Government's [Renewable Obligation] approach seeks to evaluate each technology according to need, and to provide stimulus according to the degree of that need and the potential offered by the technology."[10] This nicely echoes Karl Marx's prescription, "from each according to his abilities, to each according to his needs." Marx was of course talking about individuals rather than technologies, but the DTI's quote nicely illustrates the challenges in balancing market and regulatory forces when designing renewable policies.

U.S.—Overview

As of 2003, 15 U.S. states were establishing and/or implementing RPSs,[11] and others were considering legislation to do so. These programs use a variety of definitions, approaches, and requirements. In general, however, they reflect state-level rather than national political goals. In the U.S., state governments are typically more concerned with in-state economic development, in-state employment, and competition with other states than with global environmental issues or promoting competition for its own sake. State-level RPSs, therefore, often emphasize economic development over low price or fostering a competitive market. The U.S. state of Arizona, for example, has instituted an "environmental portfolio standard" requiring utilities to obtain 0.2 percent of their electricity from renewable sources in 2001, increasing to 1.1 percent in 2007. What's notable about this RPS is that it requires that half the renewable electricity come from solar (for example, photovoltaic) technologies. The U.S. state of Nevada has a similar solar-only requirement. Both Arizona and Nevada are in regions with very high insolation (sunlight) and see a potential business opportunity in promoting solar-based electricity generation. Similarly, Maine's RPS (discussed in more detail later) was an attempt to protect in-state biomass generators.

Most of the U.S. state RPSs are too new to draw any meaningful conclusions about how well they're working. Only those of Texas and Maine are far enough along to provide useful insight into this policy tool. So far, Texas' RPS has been astoundingly effective, while Maine's has been pretty much a flop. Here we detail these two states' RPSs, with a focus on what happened and why.

Texas' RPS

Texas had over 1,000 MW of wind capacity installed as of December 2002,[12] with the bulk of that capacity installed in the period 2000–2002.Why all this activity in Texas? Some of the credit goes to Texas' very strong wind resource and its low population density, which means few planning conflicts. More important is the Texas RPS, which required a whopping 2,000 MW of new renewables by 2009. The aggressive Texas RPS was part of that state's electricity industry restructuring legislation, passed by the Texas state legislature and signed into law by then-Governor George W. Bush in 1999.

How did Texas—a state whose economy and political system are closely tied to the fossil fuel industry—end up with an aggressively pro-renewables restructuring bill? This political conundrum can be explained in part by the scope of the bill: It mandated that the state's investor-owned utilities split into three companies (one for transmission and distribution, one for generation, and one for retail electricity sales) and established a competitive electricity market for consumers. A relatively small and largely unnoticed part of the bill established the RPS and a renewable credit trading system. Renewable advocates were certainly supportive of those components of the legislation while it worked its way through the political process, whereas potential opponents were so busy with the restructuring components of the bill, they did not have the time and resources to pay much attention to the renewables components.

Description of the Texas RPS

Texas' RPS is relatively simple. It sets a renewable generation goal, by year, in megawatts of new renewable capacity (see **Table 10-3**). Specific responsibility for meeting the goal is given to electricity retailers in the state. The portion of that goal that individual retailers must meet is determined by their fraction of total state retail sales. For example, if an individual retailer sold 10 percent of all electricity sold in the state, then that retailer would be responsible for obtaining 40 MW (10 percent of 400 MW) of its electricity in the form of qualifying renewables by 1 January 2003. Retailers failing to meet the goal face a stiff penalty—the lesser of 5 US cents or twice the renewable electricity certificate (REC, see below) value for each kilowatt-hour not provided as renewable. Texas defines renewables relatively broadly, to include biomass-based waste and landfill gas.

Table 10-3: New renewable requirements in Texas' RPS

Date	New renewable requirements, MW (cumulative)
1 January 2003	400
1 January 2005	850
1 January 2007	1,400
1 January 2009	2,000

The Texas policy is notable for two reasons. First, it includes a green certificate trading program (which it calls "renewable energy credits" [REC]). Credits are granted to operators of qualifying renewable generating facilities, at the rate of 1 REC per MWh. These credits can then be sold or traded. Second, the Texas program allows for what it calls "offsets." Suppliers can take credit for renewable facilities located at customer sites that reduce total electricity demand, such as rooftop photovoltaics.

The Resulting Wind Boom

Soon after Texas' RPS was passed into law, Texas saw an astounding boom in new renewables construction. As noted above, 1,000 MW of new wind capacity was installed as of the end of 2002. A closer look at the wind contracts signed in the state reveals an interesting anomaly: Some 160 MW of wind—17 percent of all new wind contracts in 2001—were signed by utilities that were not subject to the RPS requirements.[13] These utilities were buying wind not because they were required to by the RPS but for other reasons: to supply green power programs, to meet environmental goals, and because it was pretty darn cheap. As of late 2001, wind contracts in Texas were at less than 3 US cents per kWh.[14]

Several factors played a role in wind being available in Texas at such a low price. First is the resource: West Texas has high sustained winds, with some areas achieving annual average wind speeds of 8 meters per second or better. Second is the contracting: Because of the RPS, obligated retailers are willing to sign long-term (10- to 25-year) contracts for wind power. This in turn reduces risk for wind developers, giving them access to lower-cost capital. Third is the federal production tax credit, which was at 1.7 US cents per kWh in 2001. Fourth, transmission access rules in Texas are written so as to ease wind's entry into the grid. Although the RPS doesn't get *all* the credit for Texas' low wind price and large number of new installations, it seems to have played a major role by inducing retailers to commit to long-term contracts and thereby reducing business risk for investors.

Maine's RPS

Maine's 1997 restructuring legislation included a provision that competitive electricity providers had to obtain at least 30 percent of their electricity from renewable generation. This 30 percent was much higher than any other state or country's requirements. A closer look at the Maine electricity market, however, reveals that, at the time the legislation was passed, 50 percent of Maine's electricity *already* came from renewables—25 percent from biomass, 17 percent from hydropower, and 8 percent from municipal solid waste.[15] Why then was an RPS passed with a requirement much *lower* than the actual renewable generation? The answer nicely demonstrates how not to use an RPS: Maine wanted to protect a few specific biomass plants from price competition and tried to do so with an RPS, which turned out to be the wrong tool for the job.

Understanding why the RPS called for 30 percent renewables when actual renewable generation was 50 percent requires looking at the history of Maine's electricity market. As discussed in Chapter 8, in 1978 a federal law called the Public Utilities Regulatory Policies Act (PURPA) was passed in the U.S. PURPA was essentially a feed-in law that required utilities to pay a relatively high price for electricity from certain generating technologies, including biomass. PURPA was federal (national) legislation, but its implementation was left to the states. In the state of Maine, the utility regulators set the avoided cost—the rate PURPA required utilities to pay to biomass generators—quite high, based on expectations that the avoided fuel was oil and that oil prices would increase considerably. The end result was that the regulated utilities in Maine were required to pay a relatively high price for electricity from these biomass plants. Under this pre-restructuring regulatory system, utilities were allowed to pass these costs on to their customers. This was seen by both regulators and legislators as a desirable outcome: In-state renewable generation, with its attendant environmental, economic development, and employment benefits, was supported by what was essentially a tax on electricity.

Maine's 1997 restructuring legislation brought a rude finish to this cozy set of affairs. In putting together a plan to introduce competition into Maine's electricity system, legislators had to account for the likely result that the biomass generators would not find buyers for their higher-priced output in a competitive environment. These biomass plants became, in effect, stranded costs. State legislators had some sympathy for the policy goals that motivated the support of the biomass plants in the first place—making use of an in-state fuel resource, diversifying the energy mix, supporting an in-state industry—but maintaining the PURPA contracts was not consistent with a competitive electricity market. The 30 percent RPS was essentially a compromise between providing a truly competitive market, based entirely on price, and protecting the in-state renewable generators. In the words of one state policy maker, there was a desire not to "pull the rug out" from biomass plant owners and operators who built their plants with the expectation of continuing to receive favorable prices for their output.

The 30 percent RPS did run into an interesting political problem in 2000. In Maine, as in much of the U.S., the expectation was that restructuring would lead to lower prices. Restructuring was sold as a way to reduce electricity costs, and much of the political pressure for restructuring across the U.S. came from large industrial electricity users, who hoped for and expected significant electricity price reductions.

The reality was quite different. California, New York, and some other states saw significant electricity price *increases*, not drops, following restructuring. The reasons for this were varied and under debate. They included an unexpected rise in natural gas prices, a healthy economy that led to higher-than-expected electricity demand, and high perceived risk in building new power plants. In Maine, as well, the expected electricity price reductions did not materialize. The RPS received some blame for this. According to state regulators, "The current portfolio requirement appears to have resulted in a substantial premium on the cost of electricity in Maine, without any clearly identifiable benefits to the state."[16] This premium was estimated to be 0.1 to 0.5 US cents per kWh, for all kilowatt-hours sold in the state. (The premium paid to the biomass producers was recovered through an increase in the price of all kilowatt-hours sold.)

Due largely to its high perceived cost, there was a serious proposal in late 2000 to eliminate the RPS and replace it with a "systems benefit charge"—essentially a tax applied per kilowatt-hour, with the resulting revenue used to support renewables. After much politicking, this proposal was put aside, and as of late 2001, Maine's RPS was still in place, although just barely.

In assessing Maine's RPS experience, it's important to note that Maine's RPS was atypical for two reasons:

- It was intended to protect existing renewable resources, rather than to encourage new renewables construction.

- It was targeted at a specific technology—biomass—rather than at renewables overall. In fact, it was targeted at a small number of specific plants.

The controversy over Maine's RPS is essentially one of costs and benefits. The *costs* are fairly clear—as noted, they were estimated at 0.1 to 0.5 US cents per kWh, for all kilowatt-hours sold in the state. The *benefits*, however, are elusive. As no renewables were built, there is nothing to point to as evidence of success. And state policy makers are privately dubious that the biomass plants need the subsidy—some feel that the biomass plants could survive without the RPS.

Overall, Maine's experience with the RPS was not a success. Although it did provide some protection to existing renewable resources, it did little or nothing to promote new renewables capacity, build a competitive renewables industry, or integrate renewables into the newly competitive electricity market. What can be learned from Maine's experience?

- New renewable construction is a much more appealing and marketable policy goal than supporting existing renewable capacity.

- Unless it's very clear that an existing generating resource will close without public support, there will always be suspicion that such support is unnecessary.

- Any policy targeted at a specific technology or facility will be controversial due to the perception that it's an inappropriate subsidy. The strength of the RPS is that it results in price-based competition; but Maine's RPS was designed so that there was no such competition.

Close

Support for the RPS approach is spreading. Fifteen U.S. states and the UK have RPS-like requirements. Austria, Belgium, Italy, and Sweden have or have considered RPSs. The Netherlands has considered adding an RPS-like requirement but is waiting to see how its voluntary green market works out. Denmark did have an obligation-like system planned, but shelved it in late 2001 because of disagreements over how the associated green certificate program would work. Although the RPS approach is less market-like than, for example, voluntary green markets, its simplicity, transparency, and success at meeting its stated renewable generation goals make it an attractive option.

Notes for Chapter 10

1. See also N. Rader and S. Hempling, "The Renewables Portfolio Standard—A Practical Guide," prepared for the National Association of Regulatory Utility Commissioners (February 2001), available from www.hemplinglaw.com/writings.htm (downloaded 26 May 2002).

2. This section details the policies for England and Wales. Those for Scotland and Northern Ireland were slightly different.

3. Department of Trade and Industry, "New and Renewable Energy: The Renewables Obligation Preliminary Consultation" (undated), from www.dti.gov.uk (downloaded 19 October 2001), p. 16.

4. Department of Trade and Industry, "New and Renewable Energy: The Renewables Obligation Statutory Consultation," undated, from www.dti.gov.uk (downloaded 5 July 2002), p. 18.

5. Department of Trade and Industry, "The Renewables Obligation Preliminary Consultation," p. 3.

6. Department of Trade and Industry, "The Renewables Obligation Statutory Consultation," p. 26.

7. Department of Trade and Industry, "The Renewables Obligation Statutory Consultation," p. 23.

8. Department of Trade and Industry, "The Renewables Obligation Preliminary Consultation," p. 40.

9. As quoted in S. Boyle, "Between a ROC and a Hard Place," *Renewable Energy Report*, no. 21 (1 November 2000).

10. Department of Trade and Industry, "The Renewables Obligation Preliminary Consultation," p. 18.

11. As of December 2002, these included Arizona, California, Connecticut, Hawaii, Illinois, Iowa, Maine, Massachusetts, Minnesota, Nevada, New Jersey, New Mexico, Pennsylvania, Texas, and Wisconsin. See www.dsireusa.org.

12. American Wind Energy Association (AWEA), "Wind Energy Projects," from www.awea.org/projects/index.html (downloaded 20 May 2003).

13. R. Wiser and O. Langniss, "The Renewables Portfolio Standard in Texas: An Early Assessment," Lawrence Berkeley National Laboratory Report LBNL-49107 (November 2001), from http://eetd.lbl.gov/ea/EMP (downloaded 4 February 2002), p. 5.

14. R. Wiser and O. Langniss, "The Renewables Portfolio Standard in Texas," p. 4.

15. Data from "Case Study—Maine," from www.eren.doe.gov/state_energy/policy_casestudies_maine.cfm (downloaded 11 June 2001).

16. Maine PUC, "PUC Renewable Resource System Benefit Charge Proposed Legislation" (27 September 2000), from www.state.me.us/mpuc/2001legislation/proposed_legislation.htm (downloaded 20 June 2001).

11

Green Certificates

Green certificates (also known as tradable renewable certificates [TRCs], green tags, and renewable obligation certificates [ROCs]) are a new and very promising tool for promoting renewable electricity generation. Green certificates are essentially an accounting tool—a way to account for (and monetize) the environmental attributes of renewable-sourced electricity generation. Their set-up costs are reasonable; they require little direct subsidy or other direct payment by either government or the private sector; and they have the potential to dramatically decrease the costs of renewables. Also, as they are not a market intervention or subsidy, they fit neatly into the increasingly pro-market electricity business and are politically noncontroversial.

So what's the catch? For one, they are quite complex, both in concept and in execution. More important, it's not yet clear if they work—that is, if they do much to promote renewables (**Table 11-1**).

Table 11-1: Strengths and weaknesses of green certificates

Strengths	*Weaknesses*
Widespread political support with little or no direct political opposition	Complicated to understand and implement
Generators like them, as they result in a new revenue stream	Effects on renewables largely unknown
	Unclear relationship with carbon or other pollutant trading
Reasonable administrative costs	International trading raises difficult administrative and policy questions
A market mechanism that is economically efficient	

There is currently a flurry of activity at state, national, and international levels to establish working green certificate programs. Although there is very little real-world experience with this new and somewhat opaque concept, early results suggest that green certificates could play a significant role in the renewable electricity business.

What Is a Green Certificate?

The basic concept behind a green certificate is straightforward. A renewable electricity generator can be thought of as providing two products: the electricity into the grid and the environmental attributes associated with the renewable generation (such as reduced carbon dioxide relative to fossil-fired generation). These environmental attributes can be represented by a green certificate. This certificate can then be traded and valued on a secondary market.

How might this work? Consider, for example, a wind turbine operator. Having to bundle together the electricity and its "greenness" into one product limits the number of potential buyers. Separating them, however, both expands the number of potential buyers and simplifies the transactions. The electricity itself can be sold directly into the grid, whereas the green certificates can be sold at another time to someone not directly served by the same grid.

From the perspective of a buyer or user of green electricity, green certificates mean that one is not limited to buying green electricity from green generators located nearby. Instead, one could buy from any generator that has green certificates, regardless of location, technology, or timing of its actual generation. Or one could simply buy them from a central trading Web site.

Demand for green certificates could come from:

- A nonrenewable generator or retailer seeking to meet its renewable portfolio standard (RPS) requirement.

- A green retailer ensuring it has enough green capacity to meet its customers' needs.

- A municipality or business putting its sustainability/environmental responsibility goals into practice.

- A trader expecting the value of the green certificate to increase.

- A government trying to meet its mandated carbon or renewable generation goals.

A green certificate is essentially an accounting tool. However, a green certificate's potential for simplifying and reducing the cost of complex transactions means that it could dramatically alter the nature of the renewable generation business. By allowing for a market for the environmental attributes of renewable generation, it also creates incentives to produce renewable electricity at the lowest possible cost, as buyers will seek out the lowest-cost green certificates.

Trading Environmental Attributes: Not a New Idea

The idea of trading environmental attributes is not new. There are active markets for sulfur oxides (SO_x) and nitrogen oxides (NO_x) in the U.S., and a number of countries have pilot programs to trade carbon dioxide (CO_2) as well. An overview of the U.S. SO_2 program, which has been in place for over ten years, illustrates the potential of creating a market in environmental attributes.

The 1990 Clean Air Act Amendments (U.S.) established a tradable emissions allowance program to encourage the electricity generation industry to minimize the costs of reducing SO_2 emissions. Generators are required to hold an allowance, or certificate, for each ton of SO_2 they emit. There are a fixed number of allowances. Generators can either obtain enough allowances to cover their emissions or reduce their emissions.

Since the program began in 1995, "the experience has been full of pleasant surprises," noted one analyst.[1] Emissions have dropped below the required level, and the market price of SO_2 certificates has been much lower than expected. Although initial expectations were for prices in the range of US$750 to $1,500 per ton,[2] actual prices have generally been US$100 to $220 per ton.[3] In other words, a greater degree of environmental protection has been found at a much lower cost than expected. A more detailed evaluation of the program, which compared the results to what would have been expected from a more traditional regulatory-driven emissions control system, also found significant savings due to the trading system.[4]

The results of the SO_2 trading program are a classic example of what market forces can provide. In this case, a financial incentive was created to find better, cheaper ways to reduce SO_2 emissions, and a wealth of creativity and innovation then emerged in response to that incentive. "The [SO_2 trading] experience to date illustrates that the potential savings from incentive based environmental policies such as the allowance trading program are enormous."[5] Not everyone loves the program, of course. Some still question the ethics of selling a "license to pollute." But there is general agreement that the program clearly resulted in reduced emissions at a much lower cost than expected.

The SO_2 example and the nascent NO_x and CO_2 trading markets are not perfect analogies to green certificates. They involve trading the "bads" (SO_x, NO_x, CO_2) rather than the "goods" (environmental benefits of renewable generation); they are cap-and-trade programs (meaning they involve trade of a set number of certificates); and they offer a choice between buying the appropriate certificate or reducing emissions. Green certificates, in contrast, involve trading an undefined number of certificates with, depending on the application, no alternative to buying a certificate. These examples do, however, illustrate that it is possible to have an active and successful secondary market in environmental attributes.

History of Green Certificates: The Netherlands' Green Labels, 1998–2000[6]

There is a bit of real-world experience with green certificates. The first green certificate trading system began in January 1998 in the Netherlands. Interestingly, this effort was initiated and carried out by industry, not by government. In 1997, the Dutch association of electricity distribution companies EnergieNed set out an Environmental Protection Plan for 1997–2000. This plan included a commitment to deliver 1.7 terawatt-hours (TWh) of renewable generation by 2000 (roughly 3.2 percent of the total, then representing a doubling in renewable electricity generation).

The problem faced by EnergieNed was how to divide up responsibility for obtaining the new renewable electricity. Some of its member companies had little in the way of renewable resources, making it difficult for them to obtain new renewable generation. But it was felt that all companies needed to take some responsibility for meeting the 1.7-TWh goal. The answer they came up with was the Green Labels system.

Each of EnergieNed's member companies was given a renewable generation goal for 2000, which was set in proportion to 1995 kilowatt-hour sales. Companies committed to meeting the goal and to documenting having done so by handing over Green Labels representing the required amount of renewable kilowatt-hours on 1 January 2001. Producers of renewable energy received one Green Label for each 10,000 kilowatt-hours (kWh) of qualifying renewable energy delivered to the grid.[7] A Web site was established to allow member companies to trade Green Labels. This was a voluntary system, not directly involving the government or any regulatory process.

The program got off to a slow start in 1998, as trading was light and fewer than expected Green Labels were produced. This was due in part to member companies choosing to see what happened in 1998–1999, as the final reckoning was not until the end of 2000. Some administrative problems existed also, specifically with the registration of the Labels. But as the program continued and the problems were sorted out, it became clear that there was a true shortage in renewable supply. Significant new construction of wind turbines had been expected to take place (motivated in part by the Green Labels system), but local planning constraints meant that many of these projects were delayed or canceled. One company went so far as to buy Green Labels from a wind farm in the UK, even though it wasn't clear if non-Dutch generation could be counted toward a company's goal.

In the end, the distribution companies did not meet their 1.7-TWh goal by the end of 2000. They did get to about 1.3 TWh, or about 25 percent short of the goal. Many attribute this shortfall to the difficulties in finding sites for new wind turbines and point to local governments as the culprit. Some, however, point out that the entire system was voluntary and that there was no penalty or other disincentive for failing to meet the goal.

Renewable producers generally liked the Green Labels system. The organization of private wind turbine owners noted, "the trade in Green Labels has resulted in a better negotiation position for the private operator (and) a better price for wind electricity...(our) evaluation is positive."[8] Prices for wind did increase by about 0.7 US cents per kWh following the introduction of the Green Labels—although few shared the wind owners' view that this result was "positive." Prices for the Green Labels themselves fluctuated but were typically about 2 US cents per kWh (or about US$200 for a 10,000-kWh Label).

The Netherlands' Green Label system ended at the end of 2000 and is now being replaced with a new green certificate system.

Current Status of Green Certificates

The green certificate market in both the U.S. and the European Union (EU) is chaotic and in transition. Several countries and U.S. states are in the midst of implementing or planning for green certificate systems; a credible effort to establish an international green certificate system is underway as well (**Table 11-2**). These systems vary tremendously, from small companies in the U.S. selling green certificates door-to-door in competition with regular green power programs, to large sales among electricity providers in the UK made to meet renewable obligation (RO) requirements.

Table 11-2: Selected green certificate and renewable electricity trade programs

Country/area	Description	Status
Belgium	Green certificate to support meeting renewable obligation	Launched in 2002
Netherlands	Green certificate to support voluntary demand	Launched in 2001
UK	Green certificate to lower costs of renewable obligation	Launched 2001/2002
Renewable Energy Certificate System (RECS), European Union (EU)	EU-wide certificate system	Six country test in 2001
U.S., multi-state	Certificates for several different types of buyers	Nine marketers active in 2002
Texas, U.S.	Renewable Energy Credit Program to support state renewable portfolio standard	Trading began in 2002

The current chaos can best be understood by breaking down green certificate systems into two types: those serving primarily renewable portfolio standards and those serving primarily green power programs and related nonregulatory markets.

Belgium, the UK, and Texas all have established green certificate systems to reduce the cost of meeting their renewable obligation/portfolio standard. (See chapter 10 for a discussion of those requirements.) Texas, for example, established a green certificate system along with its RPS. Electricity retailers in Texas are required to obtain a certain percentage of their electricity in the form of qualifying renewables, and this percentage increases over time. To demonstrate that they are meeting this requirement, these retailers must provide the appropriate number of green certificates to the state regulators. For example, if the Acme Electricity Company's share of the Texas RPS is 1,000 megawatt-hours (MWh) in 2002, then the company must obtain 1,000 MWh worth of green certificates in 2002. Note that it is not necessary for the company to actually purchase or deliver any renewable electricity; it merely must obtain the correct number of green certificates. As of 2002, the Texas program appears to be working quite well. In its first year of operation, fifteen generators operating 870 MW of renewables (mostly wind) registered to sell certificates, while thirteen competitive retailers and fifteen traders registered to buy and trade.[9]

In contrast, green certificate systems in the Netherlands and U.S. states other than Texas are driven not by RPSs but by other demands. The Netherlands has a very successful voluntary green power program, with about one-fourth of the population buying green power (see chapter 6). Dutch electricity retailers selling green electricity obtain green certificates in proportion to their green electricity sales. Similarly, there are a number of green certificate marketers in the U.S. that buy green certificates from renewable generators and then sell them to a variety of users. These users include green electricity retailers as well as individual electricity consumers. Early results in the U.S. suggest that selling certificates in a voluntary market is problematic for two reasons. Small end users, such as residential users, don't understand the certificates, and larger users interested in renewables for their public relations and public image value are hesitant to buy certificates because they are unsure if the public will view them as legitimate.

Green certificates are even more abstract than green energy, so it's not surprising that they require a considerable education and marketing effort to be seen as legitimate. The perceived legitimacy of green certificates did receive a significant boost in 2002 when an independent nonprofit organization in the U.S. established a green certificate certification system (that is, a set of standards that the certificates must meet) and then certified several firms' green certificate offerings.

What Are the Current Challenges for Green Certificates?

Green certificates are still novel and many questions remain. The first major challenge is communicating the idea behind them. They are a new concept and, as such, will require some effort to explain. Most professionals involved in renewable energy at the *wholesale* level are familiar with green certificates; the communication and education challenge is now at the *retail* level. Utilities using green certificates to supply green pricing programs, for example, have a choice to either explain the certificates to electricity users in a way that can be quickly understood or simply sell the green tags as green energy without attempting the difficult explanation.

Double counting takes place when credit is sought or given repeatedly for the same environmental attributes so that the green characteristics are "counted" more than once. If this type of counting is allowed, it could undermine public confidence in the integrity of green certificates. Because green certificates are a new concept in jurisprudence, double counting in certain forms may not be restricted by law. Double counting gets particularly complex in situations in which green certificates are further unbundled to sell different attributes separately, such as CO_2 offsets. (This is known as disaggregation.) An effective registration and tracking process would "retire" the attributes once the green tag is sold to avoid a double sale and, hence, a double counting.

One particularly thorny situation occurs when marketers attempt to create green certificates out of green energy being sold under preexisting bilateral contracts. Utilities purchasing such energy may be assuming that they are buying its environmental attributes, only to discover that the marketer with whom they are contracting is claiming that it retained the title to the attributes—because the contract was silent on the issue—and that it can now create green certificates from that energy. Such disputes will eventually disappear as those writing contracts learn that they need to explicitly account for who gets the green certificates.

One of the most intriguing and complex applications of green certificates would be to use them for international trading of renewable energy. The economic argument for international trades of renewable energy is similar to that for international trade of any sort: Some countries, due to resource availability, surpluses, or other factors, can produce some products and services at lower cost than other countries. It is therefore more economically efficient to allow each country to produce those goods and services for which it has a competitive advantage and to import others. That argument ignores political, environmental, and equity concerns; nonetheless, the idea is an interesting one.

There have been a few such international trades of renewable energy already. A German organic food company, for example, purchased 63,000 kWh of green certificates from an Irish wind generator in 2000. The company paid about 1.4 US cents per kWh for the certificates (which, of course, did not include the electricity itself). An even more imaginative trade involving wind power from China shows the possibilities: The Dutch utility Nuon holds a majority stake in a 24-MW wind farm on the Chinese island of Nan'ao (near Shantou, in southern China). Nuon has raised the possibility of offering Chinese wind power, via green certificates, to green electricity buyers in the Netherlands, at a lower price than Dutch wind power.

Allowing for international trades of green certificates raises a tangle of difficult questions, including the following:

- *Which country gets the political credit?* Should it go to the generator (provider of the green certificates) or the user? With many EU countries scrambling to meet their renewable energy and carbon goals, the question of how to count international trades is a controversial one.

- *How should differing levels of subsidy be dealt with?* EU countries offer varying levels of public support for renewable electricity. Should a generator in a country with a high level of subsidy be allowed to sell the resulting green certificates to a country with a low subsidy?

- *How can logistical problems be resolved?* Getting different measurement, tracking, and registering systems to work together promises to be a logistical mess.

There are efforts underway to promote such international trading. One group, known as the Renewable Energy Certificate System (RECS), is working to define a format for green certificates that would be accepted across the EU and has run test trading sessions. However, it is likely that international trading of green certificates is still some years away and will need to wait until there is a critical mass of countries with their own internal green certificate trading systems. By that time, international carbon trading may make international green certificate trading irrelevant.

Other policy hurdles and practical difficulties to be overcome include:

- Complex issues of property law that could surface concerning the transfer of title of environmental attributes and the retirement of those attributes.

- Questions about how long marketers should be allowed to bank green tags before the true-up (the accounting process of matching sold green tags with generated green energy).

- Mixed response from environmentalists. Some have opposed green tags based on the belief that they will create disincentives for building new renewables in certain cases and deepen regional pollution.

Next Steps

Green certificates are a market-based strategy. They do not alter or skew the market; they simply allow it to function more efficiently. They hold the promise of inducing a marked increase in economic efficiency, and the U.S. SO_2 trading experience suggests that green certificates could sharply reduce the costs of renewable electricity. When used in conjunction with an RPS or a voluntary green market, they will likely reduce costs and improve market functioning. In addition, green certificates are politically noncontroversial, as they require little in the way of direct spending and do not favor one group over another. Finally, their costs are relatively low (relative to direct subsidies or other strategies).

But all is not rosy. Green certificate systems are cheap, but they are not free. Getting a green certificate system up and running requires an up-front investment in software, Web design, and other similar costs. The Netherlands' experience found that these costs were about 2 percent of the price paid per certificate.[10] Subsequent systems will probably see lower costs, as Web design improves and sales volumes increase. Perhaps more important, the benefits of green certificates are difficult to quantify. Experience to date suggests that they can work well in conjunction with other programs such as RPSs (as in Texas), but that direct sales of certificates to end users is problematic due to their complexity and users' lack of familiarity. At least one U.S. green energy retailer is buying green certificates, rather than green energy, to meet its customers' green energy needs and is not telling its customers it is doing so. This shortsighted and less than honest approach will likely end badly.

Green certificates are like green energy was five or so years ago—a new concept, viewed with a mixture of intrigue and suspicion, which will have to prove itself to an occasionally fickle market. Much like the U.S. SO_2 trading experience, it's likely that, in the long run, green certificates will play a valuable role in the wholesale market but will remain a retail oddity.

Notes for Chapter 11

1. D. Bohi and D. Burtraw, "SO$_2$ Allowance Trading: How Experience and Expectations Measure Up," Resources for the Future (RFF) Discussion Paper 97-24 (February 1997), from www.rff.org (downloaded 3 March 2002), p. 1.

2. D. Bohi and D. Burtraw, "SO$_2$ Allowance Trading," p. 8.

3. U.S. Environmental Protection Agency, "Monthly Average Price of Sulfur Dioxide Allowances," from www.epa.gov/airmarkets/trading/so2market/prices.html (downloaded 20 June 2002).

4. C. Carlson et al., "Sulfur Dioxide Control by Electric Utilities: What Are the Gains from Trade?" Resources for the Future (RFF) Discussion Paper 98-44-REV (April 2000), from www.rff.org (downloaded 3 March 2002).

5. D. Bohi and D. Burtraw, "SO$_2$ Allowance Trading," p. 2.

6. This discussion draws in part from G. Schaeffer et al., "Tradable Green Certificates: A New Market-Based Incentive Scheme for Renewable Energy," Energieonderzoek Centrum Nederland (ECN) Report ECN-I-99-004 (1999), from www.ecn.nl (downloaded 11 November 2001), pp. 14–17; and M. Voogt et al., "Renewable Electricity in a Liberalized Market—The Concept of Green Certificates," *Energy and Environment*, v. 11, no. 1 (January 2000), pp. 65–79.

7. Qualifying renewables included solar, wind, hydropower less than 15 MW, landfill gas, and biomass gasification. M. Voogt et al., "Renewable Electricity in a Liberalized Market—The Concept of Green Certificates," p. 72.

8. As quoted in Schaeffer et al., "Tradable Green Certificates," p. 17.

9. "Texas Renewable Energy Credit Trading Program Annual Program Summary for 2001," from www.texasrenewables.com/reports.htm (downloaded 26 June 2002).

10. R. Haas (editor), "Promotion Strategies for Electricity from Renewable Energy Sources in EU Countries," Institute of Energy Economics, Vienna University of Technology (June 2001), p. 25.

0-595-31218-7

935750

Made in the USA